Cognitive Mapping

Cognitive Mapping is a comprehensive account of all facets of cognitive mapping research. This book provides an overview of the historical genesis of the subject area, a description of the current states-of-play, and a 'map' of what future research should investigate. Each chapter is divided into three sections – 'past', 'present' and 'future'.

Topics that are covered include:

- the links between spatial behaviour and spatial decision
- learning a new environment
- spatial learning from maps and virtual environments
- learning space at different scales
- spatial learning across the life space
- the relationships between gender/visual impairment and spatial cognition

This important work brings together specially commissioned chapters, written by international academics from a variety of different disciplines, to explore all the major theoretical and empirical strands of research developed over the past forty years.

Rob Kitchin is a Lecturer in Human Geography at the National University of Ireland, Maynooth, Ireland. His research interests include cognitive geography, spatial behaviour, disability, cyberspace, and social geography.

Scott Freundschuh is an Associate Professor at the University of Minnesota, Duluth, USA. His research interests include a wide range of topics concerning maps and spatial knowledge acquisition in children and adults.

Routledge Frontiers of Cognitive Science
Series Advisor Tim Valentine

Cognitive Mapping

Past, present and future

Edited by
Rob Kitchin and
Scott Freundschuh

London and New York

First published 2000 by Routledge
11 New Fetter Lane, London EC4P 4EE

Simultaneously published in the USA and Canada
by Routledge
29 West 35th Street, New York, NY 10001

Routledge is an imprint of the Taylor & Francis Group

© 2000 Rob Kitchin and Scott Freundschuh

Typeset in Garamond by Florence Production Ltd,
Stoodleigh, Devon
Printed and bound in Great Britain by
St Edmundsbury Press, Bury St Edmunds, Suffolk

British Library Cataloguing in Publication Data
A catalogue record for this book is available from the British Library

Library of Congress Cataloging in Publication Data
Kitchin, Rob.
  Cognitive mapping: past, present, and future / Rob Kitchin and
  Scott Freundschuh.
       p.   cm.
    Includes bibliographical references and index.
  ISBN 0-415-20806-8
  1. Cognitive maps (Psychology)—Congresses.   2. Human information
processing—Congresses.   3. Spatial behavior—Congresses.
  I. Freundschuh, Scott, 1957– .   II. Title.
  BF314.K58 2000
  153.7'52–dc21   99-054176

ISBN 0–415–20806–8

Contents

Figures and tables

Figures

Tables

Contributors

Edward H. Cornell is Professor of Psychology at the University of Alberta, Canada. He studies wayfinding and lost person behaviour in natural environments.

Scott Freundschuh is currently Associate Professor at the University of Minnesota in Duluth. He joined the faculty in 1994 and teaches courses on cartography and geographic information systems, and is the Director of the Geography Department's Geographic Information Systems and Cartographic Analysis Laboratory. His research interests include a wide range of topics concerning maps and spatial knowledge acquisition in children and adults.

Tommy Gärling is Professor of Psychology at Göteborg University. His current research focuses on attitudes, judgement, and decision-making in different contexts including travel choice. He is the Director of the Research Unit of Societal and Environmental Decision Analysis in his department.

Florence Gaunet is Junior Research Fellow at LIMSI (CNRS, Orsay). Her interests focus on the linkage between different sensory modalities and spatial representation. For several years in Marseille, she studied the role of vision in spatial memory, testing blind and sighted subjects. At the University of California and the College de France, she investigated the contribution of visual and gravitational cues to spatial memory. She is now exploring the role of language and vision in navigation.

Reginald G. Golledge is Professor of Geography and Director of the Research Unit on Spatial Cognition and Choice at the University of California, Santa Barbara. His research interests include behavioural geography, spatial cognition, spatial knowledge acquisition, cognitive mapping, human navigation with and without sight, behavioural travel modelling, spatial decision-making and choice behaviour, gender and spatial abilities, and geography and disability.

C. Donald Heth is Professor of Psychology at the University of Alberta, Canada. He studies wayfinding and lost person behaviour in natural environments.

K.C. Kirasic is Associate Professor of Psychology at the University of South Carolina. She is a life-span developmental psychologist with interests in cognitive ageing, particularly in the area of spatial knowledge and behaviour in elderly adults. She is also involved in the investigation of attachment to place in the older adult.

Rob Kitchin is Lecturer in Human Geography, Department of Geography, National University of Ireland, Maynooth. His research covers a broad range of interests including cognitive geography, spatial behaviour, disability, cyberspace, and social geography.

Robert Lloyd is Professor of Geography and Research Director for the Center of Excellence for Geographic Education at the University of South Carolina. His research interests include spatial cognition, learning theory, and cartography.

Jack Loomis is Professor of Psychology at the University of California, Santa Barbara. His research deals with complex spatial behaviour, visual and auditory space perception, visual control of locomotion, human navigation, spatial cognition, and development of a navigation system for the blind.

Patrick Péruch is Senior Research Fellow at the Center for Research in Cognitive Neuroscience (CNRS, Marseille). His work is concerned with human spatial perception, cognition, and behaviour in large- and medium-scale spaces. In particular he investigates the respective roles of the different categories of information (visual and movement-related) in human navigation, both in natural and virtual environments.

Carole M. Self is an anthropologist and full-time Senior Administrator in the College of Letters and Science at the University of California, Santa Barbara. Her research, joint with Reginald Golledge, focuses on sex, gender, spatial abilities, and spatial performance.

Lisa S. Tan is a doctoral student in the cognitive psychology programme at Northwestern University. She received her BA in Psychology from Washington University in St Louis in 1997. Her interests are in spatial cognition and the development of children's understanding of maps.

Holly A. Taylor is Associate Professor of Psychology at Tufts University. Her research addresses the link between language and other aspects of cognition, including spatial mental models, temporal reasoning, and text comprehension.

Catherine Thinus-Blanc is Director of Research at the Center for Research in Cognitive Neuroscience (CNRS, Marseille). She works in the field of spatial cognition in animals and man, at the behavioural and brain levels. Her current studies deal specifically with geometric versus local spatial encoding from a comparative perspective.

Barbara Tversky is Professor of Psychology at Stanford University. She has general interests in memory, thinking, language, and categorization that have led to research on, among other things, cognitive maps, memory distortions, eyewitness memory, imagery, relations between language and perception, events, bodies, graphic productions, and human–computer interaction.

Simon Ungar is Lecturer in Psychology at London Guildhall University. His research interests are in environmental and developmental psychology and in particular in the use of tactile maps by blind and visually impaired people.

David H. Uttal is Associate Professor of Psychology and Education at Northwestern University. He received his PhD from the University of Michigan in 1989 in Developmental Psychology. His research interests are in spatial cognition, map use, symbolic reasoning, and children's understanding of mathematics.

1 Cognitive mapping

Rob Kitchin and Scott Freundschuh

> Cognitive mapping is a process composed of a series of psychological trans-
> formations by which an individual acquires, stores, recalls, and decodes
> information about the relative locations and attributes of the phenomena
> in his everyday spatial environment.
>
> (Downs and Stea, 1973: 7)

Introducing cognitive mapping

Cognitive mapping concerns how we think about space, and how those
thoughts are used and reflected in human spatial behaviour (Downs and
Stea, 1973). Like all animals, humans move in and through space: from
toddling around the house to negotiating the playground and the neighbourhood; from travelling to work, planning a shopping excursion, to
visiting relatives in a distant city. Spatial behaviour is thus central to our
everyday lives. In order to traverse space we make hundreds of complex
spatial choices and decisions, in most cases without any reference to sources
such as maps, instead relying on our knowledge of where places are. The
necessity of understanding space is reflected in human activities that attempt
to communicate spatial information effectively to people in spaces with
which they are unfamiliar. For example, over thousands of years, humans
have employed cave paintings, three-dimensional models, maps constructed
of various natural materials (sticks, shells, rocks, sand) and man-made
materials (paper, mylar, ink), satellite images, computer-generated worlds,
and virtual environments to communicate spatial relations (Jacobson, 1999;
Blaut, 1999).

Cognitive mapping research seeks to comprehend how we come to understand spatial relations gained through both primary experience and secondary
media (e.g., maps). In other words, how people learn, process and use spatial
information that relates to the environment that surrounds them. There are
a number of research questions of practical importance that are of interest
to explorations of cognitive mapping. Some of the more obvious questions
include 'how are new routes learned?', 'how are routes between two locations remembered?', 'how are distance and direction estimates made?', 'how

are relative locations between places learned?', 'how are maps interpreted and used for location and navigation tasks?', and 'how is spatial knowledge related to spatial behaviour?'. Cognitive mapping, then, is a component of the broader field of spatial cognition research – research that concerns the understanding of spatial thought *per se*. Spatial cognition, in turn, is an important sub-part of cognition *per se*, and its understanding is thought to be central to understanding 'learning and knowing' in general (Johnson, 1987; Lakoff, 1987).

As detailed in the literature (for example, see Kitchin 1994), the terms 'cognitive mapping' and 'cognitive map' are not without their problems. The term 'cognitive mapping' has been used in three different ways. First as a descriptive title for the field of study that investigates how people learn, remember and process spatial information about an environment. Second, it has been used as a descriptive phrase for the process of thinking about spatial relations. Third, it has been used as a descriptive name for a methodological approach to understanding cognition in general, consisting of the construction of 'maps' of cognitive processes (e.g., Swan and Newell, 1994). Similarly, the term 'cognitive map' has been used in a variety ways, nearly all relating to a person's knowledge of spatial relations. Tolman (1948), the originator of the term 'cognitive map', hypothesized that we construct a map-like representation (i.e., a cognitive map) within the 'black box' of the nervous system, which is then used to guide our everyday move-ments. In this instance, the representation is structured in the same way as a cartographic map. The implication, then, is that the cognitive map acquires Euclidean properties with repeated environmental experience. So, 'cognitive' map was used by Tolman as an *explicit statement*, with spatial knowledge functionally and representationally equivalent to a map (also see O'Keefe and Nadel, 1978). The term has also been used as an *analogical device* where spatial knowledge is assumed to be like a cartographic map (Levine *et al.*, 1982, Kuipers, 1982); and as a *metaphorical device* to label spatial knowledge as functionally equivalent to a map – we act as if we have a map-like representation in our minds, although it is acknowledged that here 'map' is a hypothetical construct (Moore and Golledge, 1976; Newcombe, 1985; Siegel and Cousins, 1985). Finally, in some cases the term 'cognitive map' has been used as a descriptive term for a conceptual drawing of an individual's cognitive processes and is the outcome of the methodological process of cognitive mapping (third definition above). In this book, we use the term 'cognitive mapping' as a descriptive title for the field of study concerned with understanding spatial thought and 'cogni-tive map' to denote a person's spatial knowledge of the environment regardless of form.

Surprisingly, cognitive mapping research is a relatively recent endeavour. Whilst there were isolated studies such as Trowbridge (1913) and Hardy (1939), the vast majority of research has taken place over the course of the past forty years. This research, initiated by the seminal work of Kevin Lynch

(1960), is unusual in that it has always been multidisciplinary (although not always interdisciplinary) in character. Although psychology, particularly cognitive and developmental psychology, has tended to dominate theoretical and empirical studies, other disciplines, notably geography, but also planning, architecture, anthropology, computer science, information science, cognitive science and neuro-psychology have made significant contributions.

A reflection of its multidisciplinary character is that just five years after the publication of Lynch's book, representatives from ten separate disciplines met at the 1965 Association of American Geographers' conference in Columbus, Ohio, to present research concerning the links between spatial thought and spatial behaviour. This meeting ultimately led to the establishment of the Environmental Design & Research Association (EDRA) and to the beginning of the cross-disciplinary journal, *Environment and Behavior* (1969) (Kitchin *et al.*, 1997). These links continued to develop in the early 1970s leading to the publication of a number of edited collections with a cross-disciplinary range of authors (e.g., Downs and Stea, 1973, 1977; Moore and Golledge, 1976). However, it is fair to state that at this time links were more usually forged through disciplines approaching psychology, and using psychological thought and techniques, than vice versa (Gärling and Golledge, 1993). During the late 1970s, links started to unravel — mainly due to crises in confidence in behavioural geography (under which most geographic, cognitive mapping research was conducted). Behavioural geography was criticized by humanists for being mechanistic and dehumanizing, and by structuralists for failing to acknowledge the broader social and cultural context in which decision-making operated (namely capital relations). While some links remained in place, notably through the work of Golledge, it was not until the early 1990s that multidisciplinary links were extensively re-forged, leading to range of interdisciplinary research, notably between geographers and psychologists. Instrumental in this process has been the National Science Foundation funded National Center for Geographic Information and Analysis (NCGIA). During the ten years of the NCGIA's existence, the Center has sponsored and funded cross-disciplinary research through its many research initiatives including Languages of Spatial Relations, Spatial-Temporal Reasoning, Multiple Representations, User-Interface Design, Common-Sense Geographic Reasoning, Scale and Cognition in Geographic Space, Dynamic Representations, and Multiple Modalities and Multiple Reference Frames for Spatial Knowledge. Many of the authors in this collection have taken part in these initiatives, and are now engaged in interdisciplinary research.

Theoretically and empirically, cognitive mapping research has come a long way in the last forty years, and a number of core sub-fields are discernible — some of these are reflected in the chapter titles in this book. Throughout this period a number of specific theories relating to processes of spatial thought have been developed, tested and refined. For example, there has been a long running debate as to how people learn a new environment (see Chapter 5). Some researchers suggest that a landmark-based strategy is used in which distinctive features in the environment can be used to provide cognitive cues around which other information is anchored (e.g., Couclelis *et al.*, 1987). Others favour a path-based strategy of learning (e.g., Gärling *et al.*, 1981) or a process of recognizing vistas (Cornell and Hay, 1984; see Chapter 5). Similar debates relate to: levels and development of spatial knowledge (see Chapter 3); how people make spatial decisions (see Chapter 4); how people acquire spatial knowledge from secondary media such as maps (see Chapter 6) and virtual reality (see Chapter 7); and in small- and large-scale spaces (see Chapter 8); how knowledge develops across the life-span, particularly in early childhood (see Chapter 9) and old age (see Chapter 10); knowledge form (whether knowledge is held in imaginal or propositional form) and what language can reveal about spatial knowledge (see Chapter 11); knowledge structure (whether knowledge is structured hierarchically or in a network), brain location (where knowledge is stored within the brain structure – hippocampus, parietal cortex), and the influence of certain intervening factors such as gender (see Chapter 12) and visual impairment (see Chapter 13).

Although certain models and associated underlying theories have gained more empirical and academic support, it is fair to say that there is still a plurality of explanations as to how we come to know, process and use spatial relations relating to the environment that surrounds us.

Why undertake cognitive mapping research?

There are a number of both theoretical and practical reasons for conducting cognitive mapping research. At a theoretical level, cognitive mapping research seeks to explain fundamental questions concerning spatial knowledge acquisition, spatial processing, and how spatial understanding is realized, and spatial choices and decisions are made. At a basic level, it is commonly understood that spatial behaviour – daily navigation, such as routes chosen (wayfinding), and also decisions concerning everyday activities such as where to shop (see Chapter 4) – is predicated upon levels and use of spatial knowledge. Therefore an understanding of spatial knowledge necessarily leads to an understanding of people's spatial behaviour. Furthermore, cognitive mapping research provides basic insights into the workings of the mind, and how we learn, process and remember spatial and non-spatial information. As such, it provides evidence that feeds into fundamental debates concerning knowledge acquisition, knowledge form and structuring, and mental processes.

At a practical level, cognitive mapping research is thought to have a number of potential applications (see Kitchin, 1994, for a full review). These applications vary from those that are more conceptual in nature (the application of ideas) to those that are more concrete. At the conceptual end of the scale it is believed that an understanding of the process of cognitive mapping will provide insights that may improve urban planning, education, and professional searches (police, rescue). Here, it is hypothesized that if we understand how people think about and interact in urban environments we can design environments that facilitate rapid learning and easy retention, thus lessening the likelihood of disorientation (Canter, 1977; Downing, 1992). Similarly, if we know how people understand geographic concepts we can produce geographic materials that are easier to comprehend, or we can determine methods to teach people to be more competent at understanding geographic material (Catling, 1978; Matthews, 1992). Moreover, if we know how people make spatial decisions about where to commit certain crimes (e.g., Canter and Larkin, 1993), or how people spatially think and behave when lost (see Chapter 5) we can determine more efficient methods of spatial search.

At the more concrete end of the scale, it is believed that cognitive mapping research can provide insights that will improve technical, geographically-based media such as cartography, geographic information systems, in-car navigation systems, orientation and navigation aids for people with visual impairments, and signage placement. For example, as noted by Lloyd (Chapter 6), it is believed that an understanding of how people comprehend and use maps may lead to the developments in cartographic presentation that will facilitate greater utility. Similarly, it is anticipated that an understanding of the most effective means of spatial communication will help facilitate the development of geographic information systems (see Egenhofer and Golledge, 1998), in-car navigation systems (e.g., Jackson, 1996) and orientation and navigation aids (e.g., Golledge *et al.*, 1998), which can be used efficiently and effectively by naive lay users (non-spatial experts).

At present, theoretically-driven research far exceeds in quantity that which is practically-operationalized (Kitchin, 1994; Jackson and Kitchin, 1998). That is to say that whilst much research is motivated by potential practical applications, this research has been slow to feed into practical developments. The situation is slowly changing, with research from cognitive cartography now starting to influence map design (see MacEachren, 1995), and research relating to visual impairment starting to feed into the design of orientation and navigation aids for blind people (see Golledge *et al.*, 1998; Jacobson and Kitchin, 1997).

This book

The origin of this book was a day-long symposium held at the 1997 Annual Meeting of the Association of American Geographers in Fort Worth, Texas. The focus of this symposium was cognitive mapping in its broadest interpretation, and its aim was two-fold. First to bring together an international gathering of researchers to exchange ideas and foster research collaborations.

Second to examine recent developments, empirical research and future directions in cognitive mapping research, attempting to gain a coherent summary
of cognitive mapping research as currently theorized and practised. The
symposium consisted of twenty-seven papers divided into five sessions, plus
a plenary panel session. The papers that were presented detailed a wide
range of empirical studies and addressed a number of substantive theoretical issues across a spectrum of topical areas:

- spatial abilities
- development
- image and narrative learning
- wayfinding
- virtual worlds
- disabilities
- representation
- route learning
- environmental learning

Papers given at this symposium focused on current research and theory
in cognitive mapping, and highlighted the multidisciplinary nature of the
field and the unique opportunities for cross-disciplinary research. The varied
disciplinary backgrounds of the participants, including behavioural and
cognitive geographers, developmental and cognitive psychologists, and
computer, cognitive and information scientists, reflect this multidisciplinary
(and increasing interdisciplinary) nature of cognitive mapping research.

Papers presented during the symposium serve, in part, as the foundation
for the chapters in this book. Rather than simply publish the papers
presented at this symposium (see Freundschuh and Kitchin, 1999), the
chapters in this text go beyond the scope of the symposium providing an
account of past, present and future research within a specific field of enquiry,
cognitive mapping. In the 'past' section of each chapter, authors provide a
brief history of the main developments leading to the present thinking and
empirical approaches. In the 'present' section, authors discuss their own
empirical research to provide concrete examples of work in progress. In the
'future' section, authors identify key concerns and questions that future
research should address. Authors were chosen because of their expertise in
relation to a particular topic, and to provide a balance of contributors from
different disciplines and geographic location. The focus of each chapter was
chosen so as to provide a comprehensive overview of the field of cognitive
mapping, taking into account the range of discernible sub-fields. After forty
years of research we felt it timely to take stock of how the field of cognitive mapping has developed, the breadth and depth of present research, and
the future questions which need to be addressed.

The resulting book is, we think, a coherent, comprehensive, accessible
account of cognitive mapping research that provides a description of the

current state of play and a 'map' of what future research should investigate. It is our hope that the book will provide a useful source book to researchers in a number of disciplines interested in cognitive mapping research and also provide impetus for future interdisciplinary collaboration.

References

Blaut, J.M. (1999) Maps and space. *The Professional Geographer*, 51, 510–15.

Canter, D. (1977) *The Psychology of Place*. London: Architectural.

—— and Larkin, P. (1993) The environmental range of serial rapists. *Journal of Environmental Psychology*, 13, 63–9.

Catling, S. (1978) Cognitive mapping exercises as a primary geographical experiences. *Teaching Geography*, 3, 120–3.

Cornell, E.H. and Hay, D.H. (1984) Children's acquisition of a route via different media. *Environment and Behavior*, 16, 627–41.

Couclelis, H., Golledge, R.G., Gale, N. and Tobler, W. (1987) Exploring the anchor-point hypothesis of spatial cognition. *Journal of Environmental Psychology*, 7, 99–122.

Downing, F. (1992) Image banks – dialogues between the past and the future. *Environment and Behavior*, 24, 441–70.

Downs, R.M. and Stea, D. (eds), (1973) *Image and Environment*. Chicago, IL: Aldine.

—— and Stea, D. (1973) Theory, in Downs, R.M. and Stea, D. (eds), *Image and Environment*, Chicago, IL: Aldine, pp. 1–7.

—— and —— (1977) *Maps in Minds: Reflections on Cognitive Mapping*. New York: Harper and Row.

Egenhofer, M. and Golledge, R. (eds), (1998) *Spatial and Temporal Reasoning in Geographic Information Systems*. Oxford: Oxford University Press.

Freundschuh, S.M. and Kitchin, R. (1999) Contemporary thought and practice in cognitive mapping research: an introduction, *The Professional Geographer* (in press).

Gärling, T., Book, A., Lindberg, E. and Nilsson, T. (1981) Memory for the spatial layout of the everyday physical environment: factors affecting rate of acquisition. *Journal of Environmental Psychology*, 1, 263–77.

—— and Golledge, R. (1993) Preface, in Gärling, T. and Golledge, R.G. (eds), *Behavior and Environment: Psychological and Geographical Approaches*. London: North Holland.

Golledge, R.G., Klatzky, R.L., Loomis, J.M., Speigle, J. and Tietz, J. (1998) A geographical information system for a GPS-based personal guidance system. *International Journal of Geographic Information Systems*, 12, 727–50.

Hardy, G. (1939) *La geographie psychologique*. Gallimard, Paris.

Jackson, P. (1996) How will route guidance information affect cognitive maps? *Journal of Navigation*, 49, 178–86.

Jackson, P. and Kitchin, R.M. (1998) Editorial: applying cognitive mapping research. *Journal of Environmental Psychology*, 18: 219–21.

Jacobson, R.D. (1999) *Learning, Reading and Communicating Space: Exploring Geographies of Blindness*. Unpublished PhD, Queen's University of Belfast.

—— and Kitchin, R.M. (1997) GIS and people with visual impairments or blindness: exploring the potential for education, orientation and navigation. *Transactions in Geographic Information Systems*, 2 (4) 315–32.

Johnson, M (1987) *The Body in the Mind*. Chicago, IL: University of Chicago Press.

Kitchin, R.M. (1994) Cognitive maps: what are they and why study them? *Journal of Environmental Psychology*, 14, 1–19.

—— Blades, M. and Golledge, R.G. (1997) Relations between psychology and geography. *Environment and Behavior*, 29, 554–73.

Kuipers, B. (1982) The 'map in the head' metaphor. *Environment and Behavior*, 14, 202–20.

Lakoff, G. (1987) *Women, Fire, and Dangerous Things: What Categories Reveal About the Mind*. Chicago, IL: University of Chicago Press.

Levine, M., Jankovic, I. and Palij, M. (1982) Principles of spatial problem solving. *Journal of Experimental Psychology: General*, 111, 157–71.

Lynch, K. (1960) *The Image of the City*. Cambridge: MA: MIT.

MacEachren, A. (1995) *How Maps Work: Representation, Visualization, and Design*. New York: Guilford.

Matthews, M.H. (1992) *Making Sense of Place*. Hemel Hempstead: Harvester Wheatsheaf.

Moore, G.T. and Golledge, R.G. (eds), (1976) *Environmental Knowing*. Stroudsberg, PA: Dowden, Hutchinson and Ross.

—— and Golledge, R.G. (1976) Environmental knowing: concepts and theories, in Moore, G.T. and Golledge, R.G. (eds), *Environmental Knowing*. Stroudsberg, PA: Dowden, Hutchinson and Ross, pp. 3–24.

Newcombe, N. (1985) Methods for the study of spatial cognition, in Cohen, R. (ed.), *The Development of Spatial Cognition*. Hillsdale, NJ: Erlbaum Lawrence, pp. 1–12.

O'Keefe, J. and Nadel, J. (1978) *The Hippocampus as a Cognitive Map*. Oxford: Clarendon Press.

Siegel, A.W. and Cousins, J.H. (1985) The symbolizing and symbolized child in the enterprise of cognitive mapping, in Cohen, R. (ed.), *The Development of Spatial Cognition*. Hillsdale, NJ: Erlbaum Lawrence.

Swan, J.A. and Newell, S. (1994) Managers' beliefs about factors affecting the adoption of technological innovation: a study using cognitive maps. *Journal of Managerial Psychology*, 9, 3–26.

Tolman, E.C. (1948) Cognitive maps in rats and men. *Psychological Review*, 55, 189–208.

Trowbridge, C.C. (1913) On fundamental methods of orientation and imaginary maps. *Science*, 38, 888–97.

2 Collecting and analysing cognitive mapping data

Rob Kitchin

Introduction

In his pioneering studies, Kevin Lynch (1960) used the technique of sketch mapping in order to elucidate cognitive map knowledge of a city by its residents. In the forty years since, a wide range of techniques informed by various theoretical frameworks have been used by researchers to try to gain a clearer understanding of how we think about and behave in space. Rather than detail the mechanics of specific data generation and analysis techniques and their merits and limitations (for overviews see Newcombe, 1985; Montello, 1991; Kitchin and Jacobson, 1997; Kitchin and Blades, forthcoming), in this chapter a chronology of techniques is outlined, with key ideas and conceptual, methodological arguments detailed. The emphasis is to detail the progression of data generation and analysis since 1960, and to raise concerns and questions that have explicitly focused on these techniques, particularly within the disciplines of geography and psychology. In the final section of the chapter, how empirical cognitive mapping research might look in the future is speculated upon, along with a description of key concepts that need to be considered in future research. It should be noted that the chapter focuses specifically on the measurement and analysis of cognitive map knowledge, that is knowledge of spatial relations in the real world, and as such pure spatial cognition tests which focus on spatial understanding *per se* (e.g., paper-and-pen tests) are not discussed.

Past

Since Lynch's work, cognitive mapping research has been a multidisciplinary endeavour. This has led to a plurality of methodological approaches. Indeed, in the two decades following the publication of *The Image of the City*, there was a strong degree of innovation as researchers sought to answer questions that up until that time had received little empirical consideration. As a consequence, wide ranging data generation and analysis techniques were developed through researchers adapting traditional techniques to apply to new situations, borrowing ideas from related disciplines, and constructing new techniques to apply to particular contexts. Driving this innovation was the combination of ideas from different disciplines. For example, psychology provided techniques for the analysis of cognition and theoretical constructs for interpreting results; geography provided techniques that were ideally suited to measuring the spatial component of data and theoretical constructs relating to spatial relations in the real-world world.

Kitchin and Blades (forthcoming) classify the tests developed in this period into two main classes: uni-dimensional and two-dimensional tests. These sets of tasks were developed simultaneously, were often used interchangeably, and progressively became more sophisticated with time. Indeed, many of the original data generation techniques were relatively crude, but their improved successors are still used today.

Uni-dimensional tests seek to uncover one-dimensional aspects of cognitive map knowledge such as distance and direction. These dimensions are thought to be representative of spatial knowledge in general, but are particularly useful for measuring levels of route (procedural) knowledge. Kitchin and Blades (forthcoming) divide uni-dimensional tests into three sub-categories: distance tasks, direction tasks, and naturalistic tasks (which were developed later).

Distance tasks assess a subject's knowledge of the distance between locations. In his review, Montello (1991) identifies five groups of tests designed to measure cognitive distance estimates: ratio scaling, interval and ordinal scaling, mapping, reproduction and route choice. All were first formulated and used between the late 1960s and late 1970s.

Ratio scaling adapts traditional psychophysical scaling techniques to a distance context, with subjects estimating the distance to a location as a ratio of some other known distance, such as an arbitary scale (e.g., Allen *et al.*, 1978; Cadwallader, 1979) or the length of a ruler (e.g., Lowery, 1973; Briggs, 1973, 1976; Day, 1976; Phipps, 1979). *Interval* and *ordinal scaling* is similar to ratio scaling but differs in the level of measurement. Paired comparison requires a respondent to decide which one of a pair of distances is longer (e.g., Biel, 1982); ranking requires respondents to rank various distances in order along the dimension of length (e.g., Kosslyn *et al.*, 1974; Allen *et al.*, 1978); rating requires respondents to assign the distance between places to a set of predetermined classes that represent relative length (e.g., Baird *et al.*, 1979); partition scales require the respondent to assign distances to classes of equal-appearing intervals of length (e.g., Cadwallader, 1979). The third group of tests, *mapping*, requires several places to be represented simultaneously, at a scale smaller than the estimated environment. Distances are measured from this map for comparison with the actual distances. *Reproduction* requires respondents to provide distance estimates at the scale of the estimated distance. *Route choice* consists of inferring judgements of cognitive distance from the choice of route an individual makes when asked to take the shortest route between two locations.

In general, ratio scaling, mapping and reproduction distance data have been analysed using linear/non-linear regression. Here, the cognitive distance estimates are regressed onto the objective distance values and the relationship between the two observed. In addition, ratio scaling and reproduction estimates, along with interval or ordinal distance data, have been analysed using multi-dimensional scaling techniques (MDS). These techniques use distance data to explore the latent structure of knowledge by assessing the data's dimensionality. They do this by constructing a two-dimensional space from one-dimensional data using a series of algorithms. In essence, they construct a 'map' showing the relationship between a number of objects.

Direction tasks assess a subject's knowledge of the direction between two locations. It is generally assessed using a strategy of pointing. Pointing involves standing at, or imaging being at, a location and pointing to another location (e.g., Hardwick *et al.*, 1976). An alternative technique involves respondents being asked to draw a line across a compass which represents the direction to a place, when the central point represents the place the direction is being estimated from (e.g., Tversky, 1981). Direction estimates have been analysed by comparing the estimates to the actual directions, often through a simple subtraction process. In other cases, a technique of projective convergence has been used to construct a 'map' from estimates by calculating where estimates to the same location but from different sites intersect.

Two-dimensional data generating techniques produce data on a single plain, for example a map. Kitchin and Blades (forthcoming) identify three categories of two-dimensional data generating techniques: graphic tasks, completion tasks, and recognition tasks.

Graphic tasks consist of variants of sketch mapping. Subjects are given a sheet of paper and are asked to draw a map of a certain location (e.g., Lynch, 1960). Variants include providing respondents with a small portion of the map to provide a scale and reference (e.g., Pocock, 1973; Kozlowski and Bryant, 1977), and teaching subjects a sketch map language where specific symbols are used to denote particular features (Beck and Wood, 1976; Wood and Beck, 1976).

Completion tasks present subjects with a certain amount of data and require them to complete a task in relation to that data. Spatial cued response tests require subjects to place locations in relation to locations that are pre-placed (e.g., Ohta, 1979; Evans *et al.*, 1980). Reconstruction tests require subjects to locate places by constructing a model that represents their relative positions (e.g., Sherman *et al.*, 1979). Cloze procedure tests are highly cued spatial completion tests that require a subject to 'fill in' a missing space (an aspatial example of which would be, 'a dog barks but a cat _____?') (e.g., Robinson, 1974; Boyle and Robinson, 1979). Burroughs and Sadalla (1979) used a similar technique called sentence frames. Respondents were required to complete a set of these frames which took the typical format of: '_____ is close to _____' and '_____ is essentially next to _____.'

Recognition tasks measure how successful subjects are at identifying spatial relationships. Iconic tests require the respondent to correctly identify features on a map or aerial photograph (e.g., Blaut and Stea, 1971). Configuration tests require a subject to correctly identify which configuration, out of several, displays the correct spatial relations (e.g., Evans *et al.*, 1980; Evans and Pezdek, 1980). Verifiable statement tests require subjects to identify whether a textual description of a spatial relationship is true or false (e.g., Wilton, 1979).

Graphic tasks have been analysed using a variety of subjective classifications. Here, individual researchers, or more commonly members of a panel, judge the sketch maps using a set of pre-determined criteria, relating to content, style, structure, development, and accuracy. Analysis of completion tests varies according to test type. Spatial cued response and reconstruction data are often analysed using bi-dimensional regression, a two-dimensional equivalent of linear regression that quantifiably assesses scale, rotation and translation differences between the actual and estimated pattern of responses (e.g., Tobler, 1978). Cloze procedure and recognition tests are analysed by constructing an accuracy score which reveals as a percentage the number of correct placements or recognitions.

Whilst these tests were innovative and were used in a large volume of research they, and the theoretical frameworks which they were used to validate, were not accepted without criticism. In the first instance, the validity and utility of some techniques were questioned. In particular, the use of sketch mapping as a technique to determine the knowledge of spatial relations within an area was questioned with critics suggesting that sketch maps are difficult to subjectively score and code, are dependent upon drawing abilities and familiarity with cartography, have content and style influenced by size of paper used for sketching, suffer from associational dependence where later additions to the sketch will be influenced by the first few elements which are drawn, and often show less information than the respondent knows (Day, 1976; Boyle and Robinson, 1979; Downs, 1985; Saarinen, 1988).

This questioning of sketch mapping led to a reappraisal of other techniques. In part, some of the criticism identified that certain researchers, notably geographers, were borrowing techniques and ideas from psychology without a full appreciation of their merits and limitations (Cullen, 1976). This led to inappropriate tests being used to measure different aspects of spatial knowledge. Moreover, within each main category of tests, for example, distance measures, a plethora of specific tests were developed, many of which were derivatives of each other. This diversity of tests, each slightly different from tests used by other researchers, made comparison between studies difficult (Gold, 1992).

In the second instance, within the discipline of geography there was widespread questioning of the connections being made with (cognitive) psychology in the area of cognitive mapping studies as well as spatial preference, choice and decision-making relating to all aspects of daily life, e.g., shopping, residential location, travel patterns. Behavioural geography, as the burgeoning field of research was labelled, was criticized by structuralists and humanists who claimed that behavioural research was mechanistic, dehumanizing, ignored the broader social and cultural context in which decision-making operated, and over-emphasized empiricism and methodology at the expense of worthwhile issues and philosophical content (see Gold, 1992). Critics further warned of the dangers of psychologism, that is, the fallacy of explaining social phenomena purely in terms of the mental characteristics of individuals (see Walmsley and Lewis, 1993) and expressed concern over the lack of theory, poor research design and applied worth. These criticisms were particularly harsh as they were often made by researchers who had themselves engaged in behavioural geography studies (e.g., Bunting and Guelke, 1979).

The outcome of these debates was a theoretical shift of some behavioural geographers towards other positions and a subsequent division of analytical behaviouralists on the one side (who continue to engage in quantitative studies of cognitive mapping) and phenomological behaviouralists on the other side. The phenomenologists called for a shift in emphasis away from trying to measure and explain spatial knowledge to a focus on the life-world of individuals and their values, morals, and 'sense of place' (see Kitchin *et al.*, 1997).

This 'crisis of confidence' concerning the utility of theoretical and methodological frameworks used in the study of cognitive map knowledge was confined mainly to the discipline of geography. As such, whilst the number of studies by geographers noticeably declined at the end of the 1970s and the beginning of the 1980s, particularly in the UK, research continued in other disciplines, notably psychology. Accompanying the loss of momentum within geography was the emergence of a degree of insularity as cross-disciplinary links, which had been particularly prevalent between geography and psychology during the 1970s (see collections by Downs and Stea, 1973; Moore and Golledge, 1976), became less common (Kitchin *et al.*, 1997). This insularity meant that during the 1980s much of the focus of study concentrated on expanding theoretical understanding and providing a wider base of empirical evidence rather than on developing innovative, methodological frameworks – this often arises when two or more disciplines communicate and collaborate. As such, although our understanding of how people think about and behave in space improved, the methodological innovation that characterized work in the previous two decades waned. In addition, it seems that the utility and validity of techniques developed earlier were largely accepted as given (although see Bryant, 1984). As research in the 1990s has shown, however, methodological issues remain a key concern.

Present

In the 1990s, cognitive mapping research has undergone a resurgence. The field of research has once again become a growing area of interest in geography, and there has been a growth of interest in psychology and other areas of study such as computer science and information science. The insularity of the 1980s has been replaced with renewed interdisciplinary links (see Kitchin *et al.*, 1997; Freundschuh and Kitchin, forthcoming). There are several reasons for this renewed interest, but key has been the development of several kinds of spatial technologies (e.g., GIS, in-car navigation systems, personal guidance systems for visually impaired people) that need to be designed for use by non-spatial experts. Government-supported programmes, such as the US National Science Foundation's National Center for Geographic Information Analysis, have sponsored initiatives designed to bring multidisciplinary researchers from around the world together to share ideas, foster collaboration and set agendas for research.

A key focus within this revitalized field has been methodological considerations. Once again researchers have been seeking innovative ways to investigate cognitive map knowledge and to improve the validity and integrity of the techniques to generate and analyse data. The latter of these tasks has formed a substantial part of my own research agenda since 1991.

I have examined the ways in which conclusions drawn from a study are influenced by the method of data generation and analysis used. In Kitchin (1996) the results of a study which compared thirteen different tests designed to measure aspects of configurational knowledge are reported. Little consistency was found between the accuracy of respondents' spatial products in comparison to the objective standard (actual locations) and in comparison to their peers. For example, it was quite possible for respondents to do very well, both in comparison to an objective standard and their peers, when undertaking a spatial cued response test but do poorly when undertaking a recognition test. It was determined that these differences were mainly due to methodological biases introduced by the nature of the test. These biases consisted mainly of spatial and locational cueing. Spatial cueing refers to the amount of spatial information provided to the respondent. For example, an exercise in which respondents are asked to locate towns and cities has high spatial cueing when many spatial cues, such as the coastline or a road network, are provided to the respondents. Locational cueing refers to the number of designated places a respondent has to locate in an exercise. High locational cueing occurs when a respondent is given a set of specific places to locate: low locational cueing occurs when the respondent has an unconstrained choice of which places to locate. Spatial cueing was found to produce more accurate spatial products by providing a spatial framework upon which respondents could 'hang' their knowledge. In contrast, locational cueing introduced random, residual error into the data sets because respondents were required to locate places with which they had varying degrees of familiarity. To further compound these effects, because the locational cueing introduced large amounts of residual error, this error effectively masked the effects of spatial cueing (see Kitchin and Fotheringham, 1998). When these biases were removed the results for individuals across tests became closer both in terms of accuracy and in relation to peers. However task demands still produced some differences across tests and respondents.

Using the same data, the effects of aggregation upon research conclusions was also investigated. Data were analysed at the disaggregate level, and using collective and individual aggregation strategies (see Kitchin and Fotheringham, 1997). Collective aggregation entails aggregating each group's raw data sets together and creating a new 'average' data set. This 'average' data set is then analysed and the result used to represent a group. Individual aggregation entails analysing all of a group's individual data sets separately and then aggregating together all of the results and creating an 'average' group result. It was found that collective aggregation can lead to erroneous comparison of individual, spatial cognition because the strategy of aggregating raw data together to produce an average data set removes variation. This can lead to an 'inflated' set of results when the group data set is analysed. Similarly, individual aggregation removes the variation in the pattern of individual results. Disaggregation avoids this pitfall as all individual cases are still identified and the differences between individuals can be noted. For example, Figure 2.1 shows the collective (CA), individual (IA), and dis-, aggregate bi-dimensional regression r-squared results from a spatial cued response test. It is clear that the members of this group had a diversity in knowledge which the collective aggregate method clearly 'inflated' (CA: $r^2 = 0.72$; IA: $r^2 = 0.48$).

The results from this study indicated that test and analysis design are of critical importance, with different tests designed to measure the same knowledge producing significantly different results. As such, slight changes in the task demands of a test can lead to alternative conclusions about that respondent's spatial knowledge. Such findings have led to some concern about whether cognitive mapping data collection techniques are capable of generating data reflective of anything but the ability of the mapper to cope with the task set (Wood and Beck, 1976) or the mode of presentation (Spencer and Darvizeh, 1981). As Boyle and Robinson (1979) recognized, the methodology used to elicit knowledge enforces a medium of communication upon the respondent which may be complicated by the individual's interpretation of how this medium can be utilized:

> it is important that we recognize the artificiality of our demands and that we are aware of the probable lack of congruence between the narrow, corseted response which we require and the less formal, more flexible, and complex structures that people actually use.

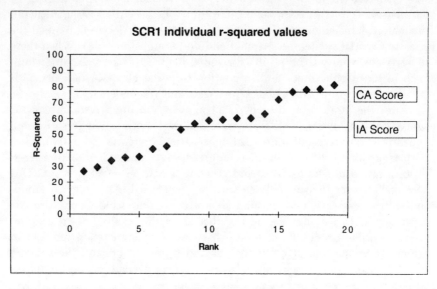

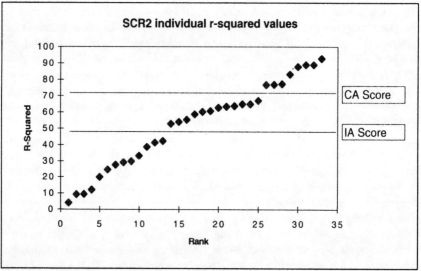

Figure 2.1 The effects of aggregation.

Indeed, the results indicate that researchers need to understand the ways in which the tests and strategies of analysis they use can introduce method-ological and analytical bias into findings. These biases need to be carefully calculated so that their effects can be compensated for when interpreting results. In the case of aggregation, given the availability of computing power results should be analysed at an individual level, or a disaggregate level for group comparisons, and should only be aggregated in specific instances, such as when computing place cognition (see Kitchin and Fotheringham, 1997).

There is evidence to suggest that these sorts of issues are being addressed as research design in the 1990s has developed significantly beyond 'simple' cognitive tasks that were often undertaken in isolation. Techniques have developed over time, with additional clauses introduced in an effort to improve the ability of particular techniques to measure specific aspects of cognitive map knowledge. In addition, the context in which these techniques have been used has become more sophisticated as experimental design has improved. Designs now usually consist of a battery of tests, designed to cross-validate findings. As noted, this strategy of multiple testing ensures a level of construct validity (whether the tests are measuring what they are supposed to) through an examination of convergent validity (tests designed to measure the same phenomenon produce the same results).

The ecological validity of laboratory settings has been questioned for a long time (see Bronfenbrenner, 1979). That is, it has been questioned whether the results achieved in a controlled laboratory have application in a real world that is more complex and variable. Increasingly experiments are taking place in real world settings rather than in a laboratory setting. This has particularly been the case with recent studies of how people with severe visual impairment learn a new route (e.g., Espinosa *et al.*, 1998; Jacobson *et al.*, forthcoming). Within these naturalistic studies, testing takes place at the scale of learning and within the actual environment the cognition of which is being assessed. Testing at this scale is thought to remove any laboratory-installed methodological biases and represent a 'truer' insight into cognition and its link to spatial behaviour. Here, instead of behaviour being inferred from knowledge, the level and nature of a person's cognitive map knowledge and abilities are inferred from both actual behaviour and evidence of knowledge gained through supporting tests.

In addition, there has been an increasing use of qualitative methodologies, both using a scientific and interpretative approach, to investigate spatial knowledge. Within a scientific approach, qualitative data is generated within very controlled tasks and usually analysed using quantitative methods. Tasks usually require respondents to describe a route or layout verbally (e.g., Taylor and Tversky, 1992). Interpretative approaches, rather than testing individuals' knowledge and abilities within controlled environments, suggest that just as much can be learnt about how people know and think about geographic space from talking to and observing individuals as they interact with an environment. Typically an ethnographic approach might be adopted. Here, the goal is to grasp the participant's point of view or relation to life and to realize their vision of their world. As such, subjects might be interviewed in depth or their behaviour within an environment observed in order to reveal their spatial understanding of an area. In my own work, I have combined aspects of the scientific and interpretative approach, using a system of semi talk-aloud protocols, semi-structured interviews and observation to investigate how people attempt to undertake a cognitive mapping test (Kitchin, 1997) and how visually impaired people spatially understand a locale (Kitchin, *et al.*, 1998). One particular strategy used was talk-aloud protocols where respondents described their actions or thoughts while performing a task, thus providing a running commentary.

Future

As we move into the new millennium, cognitive mapping research seems set to flourish given its wide range of potential applications and the healthy cross-disciplinary community that has developed. As witnessed at the end of the 1970s and early 1980s though, when a thriving community rapidly dissipated into insularity, cognitive mapping researchers need to work to ensure that the field of enquiry and links remains strong. I would argue that its methodological integrity is central to the success of cognitive mapping research. Without sound empirical evidence to support theoretical arguments and application in practical projects, cognitive mapping research is open to criticism. The future in relation to methodological aspects must (and if the progression is continued will) focus on a number of key design issues.

As detailed in the previous section, although data generation and analysis techniques and research design are becoming increasingly sophisticated, there is a need to continue to closely examine issues of data validity and integrity. This can principally be achieved through three strategies. The first is the use of multiple strategies of data generation and data analysis, using several techniques to gather and analyse data concerning the same or related issues. As my own work has shown, tests are susceptible to methodological bias and the use of multiple strategies allows the foibles of tests to be identified and compensated for. The second strategy is to generate and analyse data at a disaggregate level. Computing power now means that sophisticated analyses such as multi-dimensional scaling or the calculation of bi-dimensional regression results can be processed quickly and efficiently. Data can therefore be analysed at an individual and a group level. Again, as demonstrated in my own studies, the aggregation of individual data prior to analysis can remove variation and lead to group results that do not reflect individual performances. The third strategy is to use natural settings. As the research on spatial cognition and visual impairment is demonstrating, the results found in laboratory settings do not always apply to the real world (Jacobson *et al.*, forthcoming; Espinosa *et al.*, 1998). If we want to understand behaviour in real world settings then we need to test people's understanding and behaviour in these settings. A counter-argument might be that real world settings have too many intervening variables and are too open for effective testing. It is for precisely these reasons that testing needs to occur in natural environments, as it is these factors that people have to deal with in everyday spatial behaviour, not the controlled orderliness of the laboratory.

In addition, the utility of qualitative techniques to reveal spatial knowledge needs a more thorough examination. Recent studies, which have used scientific and interpretative approaches, suggest there is great potential. The scientific approaches have allowed an understanding of spatial language, a relatively under-studied area of spatial cognition, and, through the use of talk-aloud protocols, spatial cognition in action has been probed. Interpretative approaches allow us to try to understand the connections between spatial cognition and value systems through an examination of spatial behaviour. Here, there is a recognition that spatial behaviour is predicated on more than an understanding of spatial relations and how to get between locations.

There is also a need for the development of more sophisticated methodologies that can tease apart spatial knowledges to reveal processes of spatial thought. That is, rather than just reveal the levels of knowledge, reveal how such knowledge is structured and processed, how data learned through different media and at different scales is integrated and used in wayfinding practice. Although such work has started in earnest using traditional techniques much is still left to inference. Sophisticated battery tests linking spatial cognition tests, cognitive mapping tests, and behaviour in real world environments may be one approach; other approaches such as brain scans may also be of benefit.

Other forms of methodological development relate to the need for significantly more cross-cultural and cross-species research. Both types of study may reveal significant aspects of cognitive mapping and both provide methodological challenges. As some researchers, such as Jahoda (1979, 1980) have noted, cross-cultural studies often suffer because they conflate media experience with knowledge. For example, in societies with high rates of illiteracy, language-based tasks and pen-and-pencil tests will put those denied access to these media at a disadvantage (the same has been argued in relation to visual impairment – that the use of predominantly visual (or adapted) media will favour sighted subjects (e.g., Kitchin and Jacobson, 1997)). The challenge then is to find media that both groups are equally familiar with and then design suitable methodologies using these media that will adequately capture aspects of specific knowledge. In relation to cross-species research this question becomes more problematic, and it is likely that the only true comparator is actual behaviour and brain scans whilst particular activities are undertaken. The real danger is placing overly anthropomorphic assessments on spatial behaviour.

Moreover, there is a need to assess how applicable current methodologies are to measuring spatial knowledge derived from new spatial technologies such as GIS and virtual reality. There is an implicit assumption that knowledge gain through these media will mirror knowledge gained through traditional media, such as maps and the real world, and therefore can be validly measured in the same ways. Studies that have focused on spatial understanding of VR are finding that spatial knowledge development is significantly different, with spatial knowledge gained in VR inferior to those obtained from a map (Satalich, 1995; Tlauka and Wilson, 1996; Ruddle *et al.*, 1997) or the real world (Witmer *et al.*, 1996; Richardson *et al.*, in press). It may, however, be the case that spatial knowledge is not inferior but just different, with traditional tests unable to detect this difference. This is mere speculation, but nonetheless methodological integrity has to be determined in relation to new media.

Conclusion

In this chapter the development of methodologies to measure cognitive mapping has been detailed. Initially there was an explosion of different methods which, in part, was responsible for a crisis concerning the study of spatial cognition in the field in geography. This was because of the way tests were used, with little attention paid to their psychological context. The 1980s was a time of relative disciplinary insularity when innovation subsided and work concentrated on expanding theoretical understanding. The 1990s is characterized by renewed cross-disciplinary links, and testing has again become innovative and also increasingly sophisticated, addressing issues of methodological integrity. Testing procedures are now more likely to consist of batteries of tests, may include qualitative techniques, and may be set in natural environments. This is not to say that methodological procedures are perfect and a number of key questions for future research were identified. Indeed, future research needs to continue addressing issues of methodological integrity because without sound empirical evidence to support theoretical arguments the findings of studies are open to question.

References

Allen, G., Siegel, A.W. and Rosinski, R.R. (1978) The role of perceptual context in structuring spatial knowledge. *Journal of Experimental Psychology: Human Learning and Memory*, 4, 617–30.

Baird, J.C., Merrill, A.A. and Tannenbaum, J. (1979) Studies of the cognitive representation of spatial relations II: A familiar environment. *Journal of Experimental Psychology*, 108, 92–8.

Beck, R.J. and Wood, D. (1976) Comparative developmental analysis of individual and aggregated cognitive maps of London. In Moore, G.T. and Golledge, R.G. (eds), *Environmental Knowing*. Stroudsberg, PA: Dowden, Hutchinson and Ross, pp. 173–84.

Biel, A. (1982) Children's spatial representation of their neighborhood: A step towards a general spatial competence. *Journal of Environmental Psychology*, 2, 193–200.

Blaut, J.M. and Stea, D. (1971) Studies of geographic learning. *Annals of the Association of American Geographers*, 61, 387–93.

Boyle, M.J. and Robinson, M.E. (1979) Cognitive mapping and understanding. In Herbert, D.T. and Johnston, R.J. (eds), *Geography and the Urban Environment: Progress in Research and Applications*, 2. London: Wiley, pp. 59–82.

Briggs, R. (1973) Urban distance cognition. In Downs, R.M. and Stea, D. (eds), *Image and Environment*. Chicago, IL: Aldine, pp. 361–88.

―― (1976) Methodologies for the measurement of cognitive distance. In Moore, G.T and Golledge, R.G. (eds), *Environmental Knowing*, Stroudsberg, PA: Dowden, Hutchinson and Ross, pp. 325–34.

Brofenbrenner, U. (1979) *The Ecology of Human Development*. Cambridge, MA: Harvard.

Bryant, K.J. (1984) Methodological convergence as an issue within environmental cognition research, *Journal of Environmental Psychology*, 4, 43–60.

Bunting, T. and Guelke, L. (1979) Behavioral and perception geography: a critical appraisal. *Annals of the Association of American Geographers*, 69, 448–62.

Burroughs, W. and Sadalla, E. (1979) Asymmetries in distance cognition. *Geographical Analysis*, 11, 414–21.

Cadwallader, M.T. (1979) Problems in cognitive distance and their implications for cognitive mapping. *Environment and Behavior*, 11, 559–76.

Cullen, I. (1976) Human geography, regional science, and the study of individual behavior. *Environment and Planning A*, 8, 397–409.

Day, R.A. (1976) Urban distance cognition: review and contribution. *Australian Geographer*, 13, 193–200.

Downs, R.M. (1985) The representation of space: Its development in children and in cartography. In Cohen, R. (ed.), *The Development of Spatial Cognition*. Hillsdale, NJ: Erlbaum Lawrence, pp. 323–45.

―― and Stea, D. (eds), (1973) *Image and Environment*. Chicago, IL: Aldine.

Espinosa, M.A., Ungar, S., Ochaifa, E., Blades, M. and Spencer, C. (1998) Comparing methods for introducing blind and visually impaired people to unfamiliar urban environments. *Journal of Environmental Psychology*, 18, 277–88.

Evans, G.W., Fellows, J., Zorn, M. and Doty, K. (1980) Cognitive mapping and architecture. *Journal of Applied Psychology*, 65, 474–8.

―― and Pezdek, K. (1980) Cognitive mapping: knowledge of real-world distance and location information. *Journal of Experimental Psychology: Human Learning and Memory*, 6, 13–24.

Freundschuh, S. and Kitchin, R.M. (forthcoming) Cognitive mapping: Current theory and practice. *Professional Geographer*.

Gold, J.R. (1992) Image and environment: the decline of cognitive-behavioralism in human geography and grounds for regeneration. *Geoforum*, 23, 239–47.

Hardwick, D.A., McIntyre, C.W. and Pick, H.L. (1976) The content and manipulation of cognitive maps in children and adults. *Monographs of the Society for Research in Child Development*, 41, 1–55.

Jacobson, R.D., Kitchin, R.M., Golledge, R.G. and Blades, M. (forthcoming) Learning a complex urban route without sight: Comparing naturalistic versus laboratory measures. *Spatial Cognition and Computation*.

Jahoda, G. (1979) On the nature of difficulties in spatial-perceptual tasks: Ethnic and sex-differences. *British Journal of Psychology*, 70, 351–63.

―― (1980) Sex and ethnic differences on a spatial-perceptual task: some hypotheses tested. *British Journal of Psychology*, 71, 425–31.

Kitchin, R.M. (1996) Methodological convergence in cognitive mapping research:

investigating configurational knowledge. *Journal of Environmental Psychology*, 16, 163–85.

—— (1997) Exploring spatial thought. *Environment and Behavior*, 29, 123–56.

—— and Blades, M. (forthcoming) *The Cognition of Geographic Space*. Baltimore, MD: Johns Hopkins.

—— Blades, M. and Golledge, R.G. (1997) Relations between psychology and geography. *Environment and Behavior*, 29, 554–73.

—— and Fotheringham, A.S. (1997) Aggregation issues in cognitive mapping research. *Professional Geographer*, 49, 269–80.

—— and —— (1998) Spatial and location cueing effects upon cognitive mapping data. *Environment and Planning A*, 30, 2245–53.

—— and Jacobson, R.D. (1997) Techniques to collect and analyze the cognitive map knowledge of persons with visual impairment or blindness: Issues of validity. *Journal of Visual Impairment and Blindness*, 91, 393–400.

—— Jacobson, R.D., Golledge, R.G. and Blades, M. (1998) Belfast without sight: Exploring geographies of blindness. *Irish Geography*, 31: 34–46.

Kosslyn, S.M., Pick, H.L. and Farriello, C.P. (1974) Cognitive maps in children and men. *Child Development*, 45, 707–16.

Kozlowski, L.T. and Bryant, K.J. (1977) Sense of direction, spatial orientation, and cognitive maps. *Journal Experimental Psychology: Human Perception and Performance*, 3, 590–8.

Lowery, R.A. (1973) A method for analyzing distance concepts of urban residents. In Downs, R.M. and Stea, D. (eds), *Image and Environment*. Chicago, IL: Aldine, pp. 338–60.

Lynch, K. (1960) *The Image of the City*. Cambridge, MA: MIT.

Montello, D.R. (1991) The measurement of cognitive distance: Methods and construct validity. *Journal of Environmental Psychology*, 11, 101–22.

Moore, G.T. and Golledge, R.G. (eds), (1976) *Environmental Knowing*. Stroudsberg, PA: Dowden, Hutchinson and Ross.

Newcombe, N. (1985) Methods for the study of spatial cognition. In Cohen, R. (ed.), *The Development of Spatial Cognition*. Hillsdale, NJ: Erlbaum Lawrence, pp. 1–12.

Ohta, R.J. (1979) Spatial cognition and the relative effectiveness of two methods of presenting information in young and elderly adults. Unpublished doctoral thesis, University of Southern California, Los Angeles.

Phipps, A.G. (1979) Scaling problems in the cognition of urban distances. *Transactions of the Institute of British Geographers*, 4, 94–102.

Pocock, D.C.D. (1973) Environmental perception: process and product. *Tijdschrift Voor Economische en Social Geografie*, 64, 251–7.

Richardson, A.E., Montello, D.R. and Hegarty, M. (in press) Spatial knowledge acquisition from maps, and from navigation in real and virtual environments. *Memory and Cognition*.

Robinson, M.E. (1974) Cloze procedure and spatial comprehension test. *Area*, 9, 137–42.

Ruddle, R.A., Payne, S.J. and Jones, D.M. (1997) Navigating buildings in 'desktop' virtual environments: Experimental investigations using extended navigational experience. *Journal of Experimental Psychology – Applied*, 3, 143–59.

Saarinen, T.F. (1988) Centring of mental maps of the world. *National Geographic Research*, 4, 112–27.

Satalich, G.A. (1995) Navigation and wayfinding in virtual reality: Finding proper tools and cues to enhance navigation awareness. University of Washington, HIT Lab. http://www.hitl.washington.edu/publications/satalich/home.html.

Sherman, R.C., Croxton, J. and Givovanatto, J. (1979) Investigating cognitive representations of spatial relationships. *Environmental and Behavior*, 11, 209–26.

Spencer, C. and Darviezeh, Z. (1981) The case for developing a cognitive environmental psychology that does not underestimate the abilities of young children. *Journal of Environmental Psychology*, 1, 21–31.

Taylor, H.A. and Tversky, B. (1992) Description and depictions of environments. *Memory and Cognition*, 20, 483–96.

Tlauka, M. and Wilson, P.N. (1996) Orientation-free representations from navigating through a computer simulated environment. *Environment and Behavior*, 28, 647–64.

Tobler, W.R. (1978) Comparison of plane forms. *Geographical Analysis*, 10, 154–62.

Tversky, B. (1981) Distortions in memory for maps. *Cognitive Psychology*, 13, 407–33.

Walmsley, D.J. and Lewis, G. (1993) *People and Environment*. London: Longman.

Wilton, R. (1979) Knowledge of spatial relations: a specification from information used in making inferences. *Quarterly Journal of Experimental Psychology*, 31, 133–46.

Witmer, B.G., Bailey, J.H., Knerr, B.W. and Parsons, K.C. (1996) Virtual spaces and real-world places: Transfer of route knowledge. *International Journal of Human-Computer Studies*, 45, 413–28.

Wood, D. and Beck, R. (1976) Talking with environmental A: An experimental mapping language. In Moore, G.T. and Golledge, R.G. (eds), *Environmental Knowing*. Stroudsberg, PA: Dowden, Hutchinson and Ross, pp. 351–61.

3 Levels and structure of spatial knowledge

Barbara Tverksy

Introduction

In order to get from here to there without the benefit of external cognitive aids, like maps or written instructions, a traveller needs knowledge of different kinds. At a global level, the traveller needs a mental representation of an area that includes 'here' and 'there' and regions around and between them. Let's call that knowledge 'overviews'. Using that knowledge, the traveller determines a feasible route. Along the route, and especially at choice points, the traveller needs a representation of the local surroundings, with information critical to the choice highlighted. Let's call that knowledge 'views'. At yet a finer level, the traveller needs to know how to take each step or each turn of the wheel, maintaining course while avoiding pitfalls and obstacles. We'll refer to that detailed knowledge as 'actions'. Each of these levels, and they are by no means all, calls up different sorts of information from the world and the mind and joins them in ways judged appropriate to the task at hand (for analyses and reviews of mental geographic spaces see Freundschuh and Egenhofer, 1997; Mark, 1992; Montello, 1993; Tversky *et al.*, in press). Those engaged in making a robot into a competent traveller have discerned these levels of information, and more (e.g., Chown *et al.*, 1995; Gopal *et al.*, 1989; Kuipers and Levitt, 1988).

On the one hand, wayfinding is an everyday task, essential to survival, that has been accomplished by people since they evolved and by other organisms before that, using their eyes and bodies and minds. In people, language adds yet another layer of information about space. On the other hand, this apparently simple activity has presented serious challenges to the design of robots, and those challenges have elucidated the many aspects of knowledge and skill that must be combined to successfully get from here to there. Thus, psychologists, geographers, anthropologists, neuro-scientists, linguists, and computer scientists are among those who have been captivated by the study of spatial knowledge (for historical reviews see Mark *et al.*, in press; Tversky, in press).

Of course wayfinding is not the only use to which spatial knowledge is put. We use spatial knowledge to understand history and politics, to decide where to live and visit, and to make sense of the natural world around us, such as weather, stars, rocks, rivers, plants, and animals – anything that is distributed spatially in the world. Thus, the information about the spatial world that we encode and remember should be general and varied enough to serve purposes both known and not yet known. Spatial knowledge is diverse, complex, and multi-modal, as are the situations in which it is used.

In what follows, I will first discuss evidence characterizing mental representations at each of the three levels: overview, view, and action. Then I will consider the special case of language, especially with respect to overview descriptions and route directions.

Overview level: cognitive maps

The endearing yet controversial notion 'cognitive map' was coined by Tolman (1948) to declare that through experience traversing mazes in search of food, even hungry rats learned the general configuration of the mazes over and above specific routes to rewards. How to characterize a cognitive map has been a source of debate (e.g., Kuipers, 1982; Tversky, 1993). All seem to agree that a cognitive map is a mental representation of an external environment. At one extreme are those who appear to believe that a cognitive map is like a map on paper (or a mental image, Kosslyn, 1980); that is, a more or less veridical, more or less metric, unified representation of the environment. At the other extreme are those who appear to believe that a cognitive map may be an *ad hoc* collection of information from different sources put together to solve a particular problem; as such it has no inherent unity and no guarantee of consistency or veridicality. The mind contains many different kinds of knowledge structures, some truer to perception, more metric, more consistent – for example, images (Kosslyn, 1980). Others may bear structural similarities to some state or process in the world, yet are categorical and more abstract, such as mental models (e.g., Gentner and Stevens, 1983; Johnson-Laird, 1983; Tversky, 1991). Still others may be closer to a hotchpotch of multi-modal information, called a 'cognitive collage' (Tversky, 1993). There is no reason to doubt that, for some people and for some environments, mental representations are more like maps on paper, and for other situations more like collages. Nevertheless, considerable evidence from various cognitive tasks that draw on environmental knowledge suggests systematic biases and distortions that do not seem to be reconcilable in a consistent map-like structure (e.g., Tversky, 1981; 1992).

All conceptions of cognitive maps recognize that, as for maps, not all the information in the environment is represented. Rather that information is schematized. Much information is left out, some information is simplified or idealized. In his influential book, Lynch (1960) put forth the components from which images of cities are constructed: landmarks, nodes, paths, edges, and regions. Even more abstractly, these can be regarded as elements, such as objects, landmarks, streets, cities, countries, intersections, depending on the scale of the representation, organized in a spatial reference frame. In cognition, elements are represented relative to each other and relative to a spatial reference frame. Each of these relations can systematically distort spatial information. Elements, then, loosely correspond to 'what' and spatial relations to 'where' in the dissociable systems in the brain (e.g., Ungerleider and Mishkin, 1982).

Other elements

Alignment

When elements are located relative to each other, they are remembered as more aligned relative to a reference frame than they actually are. Consistent with this, a majority of people reject a veridical map in favour of a distorted one in which North and South America are placed closer to one above the other, that is in closer north–south alignment, than they actually are (Tversky, 1981). Similarly, a majority of respondents reject a veridical map of the major continents of the world in favour of a map in which North America and Europe appear in greater east–west alignment. That such effects are the result of perceptual organization, akin to Gestalt proximity, is supported by identical findings for pairs of cities, artificial maps, and meaningless blobs (Tversky, 1981).

Cognitive reference points

In any actual environment, certain elements are more prominent than others, perhaps because of perceptual salience, perhaps because of functional significance. These privileged elements, typically called landmarks, serve as reference points for many less distinguished elements (e.g., Couclelis *et al.*, 1987; Shanon, 1983). They then come to organize the space around them, defining neighbourhoods. Distance estimates from ordinary buildings to landmarks are smaller than distance estimates from landmarks to ordinary buildings – as if landmarks draw ordinary buildings toward themselves (e.g. McNamara and Diwadker, 1997; Sadalla *et al.*, 1980). This asymmetry of distance estimates is a violation of metric models of space. Distance asymmetries also appear between prototypic and atypical colours (Rosch, 1975) and prototypic and atypical exemplars of abstract categories (Tversky and Gati, 1978).

Frames of reference

Frames of reference not only serve to locate and orient entities within them. They also allow integration of different spaces into a common space. There are several natural frames of reference for environments.

Hierarchical organization

One natural reference frame for cities is the states that contain them; at a finer level, a natural reference frame for buildings are the neighbourhoods that contain them. Though flat, spaces are conceived of hierarchically. Using an example immortalized as an item in 'Trivial Pursuit', Stevens and Coupe (1978) found that the relative directions of cities are distorted toward the overall directions of the states they are in. A majority of students in San Diego incorrectly indicated that San Diego was west of Reno. The distortion was obtained for other pairs of cities, and for artificial maps. Times to make direction judgements are faster when the two elements come from different geographic entities than when the elements to be compared are from the same geographic entity, state or country (e.g., Maki, 1981; Wilton, 1979). Evidence for hierarchical organization comes from distance judgements as well as direction judgements. In general, distance estimates are smaller between pairs of elements within a geographic entity than between elements located in different geographic entities (Allen and Kirasic, 1985; Hirtle and Jonides, 1985). Comparable errors appear for small spaces devoid of meaning (e.g., Huttenlocher *et al.*, 1991; Lansdale, 1998).

Canonical axes

Another common reference frame for environments are the north–south east–west axes of the world. These distort the directions of the elements within them, as orientation of elements is often conveniently remembered relative to the orientation of an axis of the frame of reference. For example, people will make a map of South America 'upright' when placing it into a set of north–south east–west axes (Tversky, 1981). In its natural orientation, South America appears tilted relative to the canonical axes. Similar errors have been shown in judgements of directions between pairs of cities in the world, in memory for directions of roads, in memory for artificial maps, and in memory for meaningless blobs (Tversky, 1981) and has been replicated in other situations (Glicksohn, 1994; Lloyd and Heivly, 1987).

Other distortions and biases

Perspective

It is well known in vision that nearby distances are easier to discern and therefore prone to exaggeration than faraway distances. Interestingly, this bias appears in mental spaces as well. Students were asked to imagine themselves either on the east coast or the west coast. Then they were asked to judge the relative distances between pairs of cities scattered more or less evenly across the US, east to west. Students overestimated the distances of pairs of nearby cities relative to pairs of faraway cities (Holyoak and Mah,

1982). This occurred for both perspectives, suggesting that for mental spaces, perspective choice is flexible.

Distances

As we have seen, distance estimates are judged to be relatively smaller for pairs of places located in the different geographic entity than for pairs of places located in the different geographic entities. Short distances are over-estimated relative to long ones (e.g., Lloyd, 1989), a general finding in estimates of quantity (e.g., Poulton, 1989). Distance estimates are often exaggerated when there are barriers along a route (e.g., Kosslyn *et al.*, 1974; Newcombe and Liben, 1982), when there is clutter from increasing objects (Thorndyke, 1981), turns (Sadalla and Magel, 1980), number of nodes (Sadalla and Staplin, 1980b), and amount of information retained from the environment (Sadalla and Staplin, 1980a). Positive effect, on the other hand, seems to shorten distance estimates (Briggs, 1973; Golledge and Zannaras, 1973).

Simplifications

Curves are often remembered as straighter than they actually are, whether rivers, the Seine by residents in Paris (Milgram and Jodelet, 1976) or roads, by experienced taxicab drivers in Pittsburgh (Chase, 1983). Angles of intersections are schematized to 90 degrees (Moar and Bower, 1983). Regions are remembered as more symmetric than they actually are (Howard and Kerst, 1981; Tversky and Schiano, 1989), and the estimated areas of regions shrink in memory (Kemp, 1988; Kerst and Howard, 1978).

Views

Cognitive maps, however conceived, schematize the two-dimensional horizontal slice of the world. By contrast, views schematize vertical slices of the world. They have also been called 'You Are Here' pointers, consisting of a place description, a path description, a direction, an orientation, and a heading (Kuipers, 1978). People's memory for vertical views of places had been thought to be excellent (Shepard, 1967; Standing, 1973); however, recent research indicates that although general recognition of environments is excellent, changes in details and objects often go unnoticed, provided the general configuration remains the same (Simons, 1996). Scenes that are organized are remembered better than scenes that are unorganized, where the organization is primarily vertical, governed by gravity (Mandler *et al.*, 1977). Scenes fall into natural categories. At the highest level, indoor scenes are distinguished from outdoor scenes. At the basic level, people categorize outdoor scenes into beach scenes or city scenes or forest scenes and categorize indoor scenes into schools, restaurants or grocery stores (Tversky and Hemenway, 1983). Scenes seem to be so important to human behaviour that a region of the parahippocampal cortex appears to be dedicated to recognition of them (Brewer *et al.*, 1998; Epstein and Kanwisher, 1998). From the point of view of wayfinding or geographic information, the most important features of views are landmarks.

Space at the view level can also be experienced in three dimensions with the help of memory, as the set of objects surrounding the body. People are able to keep track of the relative positions of the objects around themselves effortlessly, even when the objects are not visible. People seem to do this by constructing a spatial mental framework from extensions of the three body axes and associating objects to the axes. The relative accessibility of objects depends on their directions from the body. Enduring properties of the body and the perceptual world appear to determine the relative accessibility of the axes. When upright, the head/foot axis is most accessible because it is an asymmetric axis of the body and aligned with gravity. The front/back axis is next, as it is also asymmetric, whereas the left/right axis is slowest because it has neither salient asymmetries nor any association with an environment axis (Franklin and Tversky, 1990). A variant of this analysis accounts for memory retrieval times for objects in three-dimensional displays in front of observer, with the same ordering of axis accessibility (Bryant *et al.*, 1992).

Action level

Although information about elements and spatial relations is important at overview, view, and action levels alike, somatosensory and vestibular information is especially important for the action level. Rieser (1989) has compared spatial relation judgements after real and imagined movements. In his task, participants learn a room-sized environment from observation at a particular viewpoint. Then they are blindfolded. One group is led to a new station point and the other group is asked to imagine themselves at a new station point. From the new real or imagined location, both groups are asked to point to other objects in the room. For translated movements, the navigation and imagine groups were equally fast and accurate. However, for rotational movements, the actual navigation group performed faster and more accurately than the imagine group (see also Easton and Sholl, 1995; Presson and Montello, 1994). Actual movement seems critical for keeping track of orientation. This may be because in some imagined rotations, participants fail to keep track of their changed headings (Klatzky *et al.*, 1998; but see also Franklin and Tversky, 1990, where in a different paradigm, imaginary rotations are accurate). For translations, there is no change of heading. Other research has shown that in blindfolded participants, somatosensory and vestibular information is important for keeping track of turning motion (Berthoz *et al.*, 1995; Takei *et al.*, 1996). This sensory information may be used to infer headings and spatial relations of surrounding objects. That somatosensory and vestibular information is integrated into a schema of the environment is suggested by experiments showing that counting backwards out loud interferes with keeping track of motion (Takei *et al.*, 1997).

Learning from exploration versus maps

Consistent with the conclusion that the sensory information provided by actual exploration contributes to the formation of mental representations of environments, and especially to orientation, Thorndyke and Hayes-Roth (1982) found that people who learned a two-building complex through actual exploration were more accurate pointing to the directions of landmarks in the complex than those who learned the complex from studying a map. Explorers were not superior on all judgements; in fact, those who learned from maps were more accurate in estimating direct distances between pairs of points. Similar effects were found by Taylor *et al.* (1999), who also found that learners' goals as well as mode of learning influenced what people extracted from learning. Thus, the goal to learn the layout achieved some of the benefits of learning from a map and the goal to find efficient routes achieved some of the benefits of learning by exploration.

Space of the body

Action involves the body. Somatosensory information also seems to contribute to knowledge of another space important in navigation, the space of the body. Unlike other objects, people experience bodies from the inside as well as the outside. Because sensorimotor information is not equally distributed over the body, this view predicts that some parts of the body should be more salient than others. In accordance with this, Reed and Farah (1995) found that moving the upper body facilitated detection of upper body differences in pictures of body postures and moving the lower body facilitated detection of lower body differences in pictures of body postures. Moreover, upper body changes were detected better than lower body changes, consistent with the greater sensorimotor innervation of the upper body. Further evidence for and development of this position was provided by Morrison and Tversky (1997). In a task requiring verification of named body parts, participants were faster to verify more significant than to verify large parts, where significance was loosely indicated by relative sensorimotor innervation.

Future

Since its disparate beginnings, including the seminal work of the rat psychologist Tolman and the urban architect Lynch, to name but two, the field of cognitive maps has made fascinating progress. This progress has been the result of efforts by psychologists, geographers, computer scientists, linguists, anthropologists, and others and I have reviewed only a small fraction of the work here. Despite the progress, many fascinating problems remain unsolved, including some of the very basic issues. How do we get from actions and views to overview knowledge of the world – from our own situation-based experiences to knowledge more abstract that allows us to behave intelligently in new situations? How do we integrate different kinds of information from different barely comparable sources? How do we retrieve the right – or wrong – information that we retrieve in any particular task?

Language of space

Important as experience may be, for people language serves as an impressive surrogate for experience. Well-told stories can bring tears to our eyes, whether from happy or funny or sad descriptions. The way something is described, either by ourselves or by others, affects our interpretation of it and our memory for it (e.g., Carmichael *et al.*, 1932; Levinson, 1996; Loftus and Palmer, 1974; Schooler and Engstler-Schooler, 1991; Tversky and Marsh, in press). Language has additional advantages. It is lightweight and portable. It allows mental transport to other times and places, freeing us from our own place and time and body. The evocative properties of language are especially important for describing space. Duras evokes the sultry scene in colonial Vietnam as Hardy evokes the rolling, windswept moors of England. Spontaneous descriptions can successfully guide travellers to their destinations, even in such arcane environments as Venice (Denis *et al.*, 1998; see also Streeter *et al.*, 1985). Language is by nature categorical, though it does have devices for conveying continuous information. In any case, language is more reliable for conveying categorical spatial relations than exact ones (e.g., Leibowitz *et al.*, 1993).

We are all experts in describing space just as we are all experts in navigating space. Spatial descriptions do a number of things. Of particular concern here, they locate landmarks, typically relative to each other and to a reference frame. When complete and coherent, spatial descriptions by themselves have the power to provide an adequate representation of an environment – a spatial mental model which could be either an overview (e.g., Taylor and Tversky, 1992a) or a view (e.g., Franklin and Tversky, 1990). Route directions are a special kind of spatial description, designed to take a traveller from one point to another rather than to give an overall impression of an environment. Spatial language has been studied primarily at the levels of overview and view, though there is recent work on describing actions (e.g., Habel and Tappe, in press; Tappe and Habel, 1998; Zacks and Tversky, 1999). For the most part, such descriptions are of intentions rather than motor activities.

Styles

When (English- or Dutch-speaking) people are asked to describe environments, they adopt one of three styles (Taylor and Tversky, 1996). For small environments that can be seen from a single point of view, usually an entrance, people use a gaze tour. They take a single viewpoint in the environment, usually an entrance, and describe the landmarks, typically objects in a room, relative to one another in terms of left, right, front, and back, relative to the natural viewpoint (Ehrich and Koster, 1983; Ullmer-Ehrich, 1982). When describing larger environments that cannot be seen from a single viewpoint, people use either a survey or a route perspective or a mixture of both (Denis, 1996; Linde and Labov, 1975; Levelt, 1982a, 1982b; Taylor and Tversky, 1992a, 1996; Tversky, 1996; Tversky *et al.*, 1997). In a survey description, people take a viewpoint from above and describe landmarks relative to each other in terms of an extrinsic reference system, normally the canonical north–south east–west axes. In a route description, people take the changing viewpoint of a traveller, typically characterized as 'you', in an environment, and describe locations of landmarks relative to your current position in terms of your left, right, front, and back (Taylor and Tversky, 1992a, 1992b, 1996).

Surprisingly, people frequently mix these styles in describing an environment, often without signalling and without noticeable costs in comprehension (Taylor and Tversky, 1992a, 1996). This is in spite of the fact that perspective switches can require extra time to process during reading (Lee and Tversky, in preparation). The configuration of an environment is at least in part responsible for the choice of description style. For example, survey descriptions are relatively more popular in English for environments that contain multiple routes or that contain landmarks at multiple scales (Taylor and Tversky, 1996).

Reference frames

These three perspectives correspond to the three frames of reference distinguished by Levinson (1996). Previous analyses of perspective in language had distinguished three perspectives based on the referent, the speaker or viewer, an object, or an environment, and the terms of reference (see, for example, Buhler, 1982; Fillmore, 1975, 1982; Levelt, 1984; Miller and Johnson-Laird, 1976). These analyses led to conceptual difficulties, especially in distinguishing a 'deictic' perspective based on a speaker or viewer from an 'intrinsic' one based on an object. There didn't seem to be any principled difference between describing the location of a ball as 'in front of me' or as 'in front of the house' as long as it was clear that 'front' referred to the inherent side of the person or object.

To correct the ambiguities of the previous analyses while preserving the alliance of reference frames with reference objects, Levinson (1996) proposed a new analysis of relative, intrinsic, and absolute reference frames. A relative frame of reference is based on a person; it locates a target object relative to a reference object with respect to the person, in terms of the person's front, back, left, and right. To specify such a relation requires three terms, the person, the target object and a reference object. This is one of the cases classically called deictic, as understanding the expression depends on knowing the person's viewpoint. An intrinsic reference frame is based on an object; it locates a target object relative to a reference object with respect to that object's intrinsic front, back, left, and right, a binary relation dependent on the target and reference objects, and not dependent on knowledge of a viewpoint. It does require that the reference object have intrinsic sides, which is *not* the case for many objects, for example a ball. Because they use the same terms of reference, relative and intrinsic reference frames can require disambiguation. If I say 'my bike is left of the house' it is not clear whether I mean to my own left as I look at the house or to the house's left. Finally, an absolute reference frame is based on an environment; it locates a target object with respect to a reference object in terms of, typically, north–south–east–west (other absolute reference frames are possible, for example, seaward and inward). Like an intrinsic reference frame, it does not require knowledge of a viewpoint but unlike an intrinsic reference frame, it does not require that the reference object have intrinsic sides. However, it does depend on knowledge of the cardinal directions.

The three reference frames proposed by Levinson (1996) are idealizations. Mixed cases exist: for example, an inanimate object like an entrance can be used as if it were a viewer in a relative description, and a person can be used as a referent object in an absolute description.

Now we can map Levinson's reference frames onto the three perspectives people take in describing space. The gaze tour adopted for environments that can be viewed from a single point uses a relative frame of reference; a route description uses an intrinsic frame of reference with 'you' as reference object; and a survey description uses an absolute frame of reference. These three styles of description correspond to natural ways of experiencing environments, a gaze tour to experiencing from a single viewpoint at eye level, a route description to experiencing from travelling within an environment, and a survey description to experiencing an environment from above (Taylor and Tversky, 1996; Tversky, 1996).

Reference objects

Using a reference frame imposes a heavy cognitive demand on speakers as well as addressees. To understand and often to produce relative, or intrinsic, or absolute descriptions requires mentally taking a viewpoint and mentally computing directions from the viewpoint. Some directions are easier than others; in particular, left and right are more difficult in general than front, back, above, and below (e.g., Bryant, Tversky, and Franklin, 1992; Franklin and Tversky, 1990). Some languages do not use left and right in referring to locations of objects, even for table-top environments indoors, preferring an absolute system of reference (Levinson, 1996). A far simpler, though not always possible, way to describe locations of objects is 'near X' where X is a known landmark. Understanding 'near' does not require taking a perspective or computing directions. It does assume an environment simple enough that direction information is not needed to specify the target object. In simple situations requiring specification of one of two identical objects in spatial arrays with or without landmarks and with or without indication of cardinal directions, speakers indeed used 'near' frequently, ignoring the other information available for specifying the target object. They did so often even when it seemed to violate the principle of taking the other's perspective to reduce cognitive load. That is, they said, 'the one near me' (Tversky, Taylor, and Mainwaring, 1997). This utterance does not require taking the speaker's perspective; rather, it uses the speaker as a reference object for locating the target. Interestingly, Japanese speakers were as likely to so as American (Mainwaring *et al.*, 1999).

Route directions

Route directions have been a source of fascination as well as a laboratory for linguists, psychologists, geographers, and computer scientists, among others (e.g., Couclelis, 1996; Denis, 1997; Denis *et al.*, 1999; Freundschuh *et al.*, 1990; Gryl, 1995; Klein, 1983: Tversky and Lee, 1998; Levelt, 1989; Wunderlich and Reinelt, 1982). Route directions are a paradigmatic case for studying directions, for studying the linearization of multi-dimensional situations that language requires, for studying spatial language. Those interested in language have distinguished several stages in route directions (e.g., Couclelis, 1996; Denis, 1997; Gryl, 1995; Klein, 1983; Levelt, 1989; Wunderlich and Reinhelt, 1982). Speakers first need a primary plan that includes a mental representation of the entire area. Using that, they need to determine a route. Then, they need to segment the route and construct procedures for progressing from one segment to the next. Of course, these are the requirements necessary for the speaker alone to get from here to there. The requirement to communicate the route to someone else brings in the nuances of interpersonal communication (e.g., Clark, 1996), the challenge of turning the route into words and gestures that another will understand. The consequent description is typically interactive in two senses: within each participant, there is continual interaction between mental representations of the space and linguistic expressions; between the seeker of directions and the provider of them, there is continuous interaction to make sure the directions are understood.

Denis and his collaborators (Daniel and Denis, 1998; Denis, 1997; Denis and Briffault, 1997; Denis *et al.*, 1999) are among those who have collected a body of route directions in the field. Using this corpus, they developed procedures for generating good directions (Denis, 1997). First, they recoded the individual protocols into a standard propositional format and compiled those to create a mega-description. The mega-description was then given to judges familiar with the route, who removed superfluous items. Items included by 70 per cent of the judges were retained in what was termed a skeletal description. In actual navigation, the skeletal descriptions succeeded as well or better than individual protocols rated as good (Denis *et al.*, 1999). In fact, highly-rated protocols were similar to the skeletal descriptions.

What characterizes good route directions? Denis's empirically derived prescriptions echo and expand those derived by Wunderlich and Reinelt (1982) and Klein (1983) from linguistic analyses of their corpora, and correspond to Gricean principles (Grice, 1975). Paraphrasing Denis (1997), the essential information is a set of iterative steps:

0 – locate the listener at point of departure
1 – start the progression (usually implicit)
2 – point out a landmark
3 – reorient listener
4 – start of progression
repeat 2–3–4

The steps do not necessarily appear in separate utterances. For example, in three routes on a university campus, Daniel and Denis (1998) found that about 17 per cent of utterances prescribed an action, 36 per cent prescribed an action with respect to a landmark, 33 per cent introduced a landmark, and 12 per cent described a landmark.

Confirming the work of others (e.g., Couclelis *et al.*, 1987; Siegel and White, 1975), landmarks seemed to be selected on the basis of visibility, pertinence, distinctiveness, and permanence. They served several functions: to signal the place of action change, to locate other landmarks, and to confirm the route. Similarly, actions were of two types: changes of orientations, and continuations in the same direction. The propositions most frequently eliminated by judges as superfluous were those that said to go straight, those that referred to secondary information, and those that described landmarks.

The importance of landmarks or views of critical features in wayfinding is reinforced by research teaching routes by film segments rather than language. Heft (1996) reviews studies in which observers selected the segments of films that were 'most important for finding your way'. The important segments were transitions, that is, changes of orientation, and landmarks. In other studies, participants viewed a selected film of a route and then walked the route, three times. Participants who watched a film of transitions performed better than those who watched a film of vistas between transitions.

Comparing route directions and route maps

Directions are not the only guide to travellers. Wright *et al.* (1995) found that although wayfinders appreciate maps, informants rarely provide them. Route maps differ in character from area maps, just as route directions differ from spatial descriptions. Route maps, like route directions, include only the information deemed relevant for getting from here to there, excluding other information about the region.

How can route maps be characterized? In order to find out, Tversky and Lee (1998) collected spontaneous route maps and written directions to an off-campus fast food place from passers-by outside a campus dormitory. Both maps and directions were decomposed into segments following Denis (1997). Each segment ideally contained four kinds of information: start point, reorientation, path/progress, and end point. In examining the results, we first consider necessity followed by sufficiency of the information included in route maps and directions. Then we examine how information is schematized in both.

At least 90 per cent of those providing maps and directions added information over and above the necessary information for progressing from one segment to the next. Maps often included cardinal directions, arrows, distances, and landmarks not at reorientations. Directions included the same extra information and in addition added descriptions of landmarks and paths. The extra information seemed to be designed to assure the travellers that they were on track, and have been found by others (e.g., Denis, 1997; Gryl, 1995). Human communication is not minimalist; in order to succeed, it must contain redundancies and extra information.

To evaluate sufficiency, we examined each segment for the four types of information. All of the maps were sufficient. Technically, the directions had much information missing: 75 per cent lacked either a start or an end point and 45 per cent lacked path/progression information. Discourse frequently lacks information that can easily be inferred from world knowledge (e.g., Clark and Clark, 1977). The missing information could be inferred by invoking two assumptions: continuity and forward progression. Continuity allows the assumption that a missing start point is the same as the previous end point and that a missing end point is the same as the subsequent start point. In fact, except for initial and final points, start and end points are indistinguishable. Forward progression allows inference of the path of progression. Using these two assumptions, all but 14 per cent of the directions reached sufficiency; three of them failed to specify direction of a turn where more than one direction was possible. The medium of drawing precludes some kinds of insufficiency; for example, direction must be specified.

Route maps and route directions schematized information in similar ways (see also Tappe and Habel, 1998, for units of sketch maps). In both, start and end points tended to be landmarks and intersections. In maps, landmarks were represented as icons, usually rough geometric shapes and intersections by more or less perpendicular lines (or line pairs), often labelled with nouns as in directions. Reorientations tended to be intersections. In maps, they were accompanied by arrows in nearly half the cases. In directions, reorientations were indicated by a small set of verbs: 'turn', 'take a', 'make a', and 'go', though verbs were sometimes elided and presupposed, as in 'right on Campus Drive'. In both maps and directions, the angle of turn was unspecified (as in memory for environments, e.g., Byrne, 1979; Moar and Bower, 1983; Tversky, 1981). In maps, paths were represented by lines or line pairs, straight or curved. In directions, 'go' indicated straight paths and 'follow' indicated curved ones. Like the angle of intersection, the curvature of paths is conveyed categorically in both sketches and words.

The similarities in the ways that sketch maps and verbal directions schematize routes are striking. They suggest that the same cognitive schematization and segmentation underlie both. There are differences, of course, as well. The very nature of the medium renders maps more complete than directions. What must be made explicit in maps is inferred from discourse. Maps are also a more direct mapping of space to space, which may be why wayfinders want them.

In sum

Spatial language not only provides descriptions of environments and directions to travellers, it also provides a window on spatial cognition. The perspectives used to describe things in space, especially the mixing of perspectives, reflect the perspectives, also often mixed, of experiencing things in space (e.g., Behrmann and Tipper, 1998; Bisiach, 1993). The schematization of space in language and in sketches – what gets included and excluded, what gets simplified and how – reflects the ways the mind schematizes space (e.g., Talmy, 1983, 1988). Perhaps because of the ubiquity of spatial experience and spatial language, spatial language is used metaphorically to express concepts from evaluations, as in 'top of the heap' or 'giving the high five', and moods, as in 'feeling down', to mathematics, as in graphing, and science, as in models of atoms and molecules.

Old paths and new directions

The paths to knowledge of spatial cognition are many, both in kind and in level. The paths leading from spatial cognition are also many, to kinds and levels of behaviour. Spatial cognition is a microcosm of all cognition. It depends on sensation, perception, and memory, and it determines action. It is a consequence of individual differences on the one hand, and of social behaviour, especially language, on the other. It is at once basic to individual and collective existence, necessary for survival, and applied to broad areas of individual and collective enterprise. Spatial cognition underlies the simplest of behaviours and the most abstract imaginations of art and science.

References

Allen, G.L. and Kirasic, K.C. (1985) Effects of the cognitive organization of route knowledge on judgements of macrospatial distance. *Memory and Cognition*, 13, 218–27.

Behrmann, M. and Tipper, S.P. (1998, in press) Attention accesses multiple reference frames: Evidence from neglect. *Journal of Experimental Psychology: Human Perception and Performance*.

Berthoz, A., Israel, I., Georges-Francois, P., Grasso, R., and Tsuzuki, T. (1995) Spatial memory of body linear displacement: What is being stored? *Science*, 269, 95–8.

Bisiach, E. (1993) Mental representation in unilateral neglect and related disorders: The Twentieth Bartlett Memorial Lecture. *Quarterly Journal of Experimental Psychology*, 46A, 435–61.

Brewer, J.B., Zhao, Z., Desmond, J.E., Glover, G.H. and Gabrieli, J.D.E. (1998) Making memories: Brain activity that predicts how well visual experience will be remembered. *Science*, 281, 1185–7.

Briggs, R. (1973) Urban cognitive distance. In Downs, R. and Stea, D. (eds), *Image and Environment*. Chicago, IL: Aldine, pp. 361–88.

Bryant, D.J., Tversky, B., and Franklin, N. (1992) Internal and external spatial frameworks for representing described scenes. *Journal of Memory and Language*, 31, 74–98.

Buhler, K. (1982) The deictic field of language and deictic words. (Translation of part of 1934 book in German) in Jarvella, R.J. and Klein, E. (eds), *Speech, Place, and Action*. New York: Wiley, pp. 9–30.

Byrne, R.W. (1979) Memory for urban geography. *Quarterly Journal of Experimental Psychology*, 31, 147–54.

Carmichael, R., Hogan, H.P. and Walter, A.A. (1932) An experimental study of the effect of language on the reproduction of visually perceived forms. *Journal of Experimental Psychology*, 15, 73–86.

Chase, W.G. (1983) Spatial representations of taxi drivers. In Rogers, R. and Sloboda, J.A. (eds), *Acquisition of Symbolic Skills*. New York: Plenum, pp. 111–36.

Chown, E., Kaplan, S. and Kortenkamp, D. (1995) Prototypes, location, and associative networks (PLAN): Towards a unified theory of cognitive mapping. *Cognitive Science*, 19, 1–51.

Clark, H.H. (1996) *Understanding Language*. Cambridge: Cambridge University Press.

—— and Clark, E.V. (1977) *Psychology and Language*. New York: Harcourt Brace Jovanovich.

Couclelis, H. (1996) Verbal directions for way-finding: Space, cognition, and language. In Portugali, J. (ed.), *The Construction of Cognitive Maps*. Amsterdam: Kluwer, pp. 133–53.

——, Golledge, R.G., Gale, N. and Tobler, W. (1987) Exploring the anchorpoint hypothesis of spatial cognition. *Journal of Environmental Psychology*, 7, 99–122.

Daniel, M.-P. and Denis, M. (1998) Spatial descriptikons as navigational aids: A cognitive analysis of route directions. *Kognitionwissenschaft*, 7, 45–52.

Denis, M. (1996) Imagery and the description of spatial configurations. In De Vega, M., Intons-Peterson, M.J., Johnson-Laird, P.M., Denis, M. and Maschark, M. (eds), *Models of Visuospatial Cognition*. New York: Oxford University Press.

—— (1997) The description of routes: A cognitive approach to the production of spatial discourse. *Current Psychology of Cognition*, 16, 409–58.

—— and Briffault, X. (1997) Les aides verbales a l'orientation spatiale. In Denis, M. (ed.), *Langage et Cognition Spatiale*. Paris: Masson.

—— Pazzaglia, F., Cornoldi, C. and Bertolo, L. (1999) Spatial discourse and navigation: An analysis of route directions in the city of Venice. *Applied Cognitive Psychology*, 12, 145–74.

Easton, R.D. and Sholl, M.J. (1995) Object-array structure, frames of reference, and retrieval of spatial knowledge. *Journal of Experimental Psychology: Learning, Memory and Cognition*, 21, 483–500.

Ehrich, V. and Koster, C. (1983) Discourse organization and sentence form: The structure of room descriptions in Dutch. *Discourse Processes*, 6, 169–95.

Epstein, R. and Kanwisher, N. (1998) A cortical representation of the local visual environment. *Nature*, 392, 599–601.

Fillmore, C. (1975) *Santa Cruz Lectures on Deixis*. Bloomington, IN: Indiana University Linguistics Club.

—— (1982) Toward a descriptive framework for spatial deixis. In Jarvella, R.J. and Klein, W. (eds), *Speech, Place and Action*. London: Wiley, pp. 31–59.

Franklin, N. and Tversky, B. (1990) Searching imagined environments. *Journal of Experimental Psychology: General*, 119, 63–76.

Freundschuh, S.M. and Egenhofer, M.J. (1997) Human conceptions of spaces: Implications for geographic information systems. *Transactions in GIS*, 2, 361–75.

—— Mark, D.M., Gopal, S., Gould, M. and Couclelis, H. (1990) Verbal directions for way-finding: Implications for navigation and geographic information and analysis systems. In *Proceedings, Fourth International Symposium on Spatial Data Handling*, Zurich, Switzerland, vol. 1, 478–87.

Gentner, D. and Stevens, A.L. (1983) *Mental Models*. Hillsdale, N.J: Erlbaum Lawrence.

Glicksohn, J. (1994) Rotation, orientation, and cognitive mapping. *American Journal of Psychology*, 107, 39–51.

Golledge, R.G. and Zannaras, G. (1973) Cognitive approaches to the analysis of human spatial behavior. In Ittelson, W.H. (ed.), *Environment and Cognition*. New York: Seminar, pp. 59–94.

Gopal, S., Klatzky, R.L. and Smith, T.R. (1989) NAVIGATOR: A psychologically based model of environmental learning through navigation. *Journal of Environmental Psychology*, 9, 309–31.

Grice, H.P. (1975) Logic and conversation. In Cole, P. and Morgan, J.L. (eds), *Syntax and Semantics 9: Pragmatics*. New York: Academic Press, pp. 113–27.

Gryl, A. (1995) Analyse et modelisation des processus discursifs mis en oeuvre dans la description d'itineraires. Notes et documents LIMSI No. 95–30. LIMSI-CNRS, BP 133, F-91403, Orsay, France.

Habel, C. and Tappe, H. (in press) Processes of segmentation and linearization in describing events (manuscript). In Stutterheim, C.V. and Meyer-Klabunde, R. (eds), *Processes in Language Production*.

Heft, H. (1996) The ecological approach to navigation: A Gibsonian perspective. In Portugali, J. (ed.), *The Construction of Cognitive Maps*. Netherlands: Kluwer, pp. 105–32.

Hirtle, S. and Jonides, J. (1985) Evidence of hierarchies in cognitive maps. *Memory and Cognition*, 13, 208–17.

Holyoak, K.J. and Mah, W.A. (1982) Cognitive reference points in judgements of symbolic magnitude. *Cognitive Psychology*, 14, 328–52.

Howard, J.H. and Kerst, S.M. (1981) Memory and perception of cartographic information for familiar and unfamiliar environments. *Human Factors*, 23, 495–503.

Huttenlocher, J., Hedges, L.V. and Duncan, S. (1991) Categories and particulars: Prototype effects in estimating spatial location. *Psychological Review*, 98, 352–76.

Johnson-Laird, P.N. (1983) *Mental Models*. Cambridge, MA: Harvard.

Kemp, S. (1988) Memorial psychophysics for visual area: The effect of retention interval. *Memory and Cognition*, 16, 431–6.

Kerst, S.M. and Howard, J.H. (1978) Memory psychophysics for visual area and length. *Memory and Cognition*, 6, 327–35.

Klatzky, R.L., Loomis, J.M., Beall, A.C., Chance, S.S. and Golledge, R.G. (1998) Updating an egocentric spatial representation during real, imagined, and virtual locomotion. *Psychological Science*, vol. 9, 293–8.

Klein, W. (1983) Deixis and spatial orientation in route directions. In Pick, H.L. Jr. and Acredolo, L.P. (eds), *Spatial Orientation: Theory, Research, and Application*. New York: Plenum, pp. 283–311.

Kosslyn, S.M. (1980) *Image and Mind*. Cambridge, MA: Harvard.

—— Pick, H.L. and Fariello, G.R. (1974) Cognitive maps in children and men. *Child Development*, 45, 707–16.

Kuipers, B.J. (1978) Modelling spatial knowledge. *Cognitive Science*, 2, 129–53.

—— (1982) The 'map in the head' metaphor. *Environment and Behavior*, 14, 202–20.

—— and Levitt, T.S. (1988) Navigation and mapping in large-scale space. *AI Magazine*, 9, 25–43.

Lansdale, M.W. (1998) Modeling memory for absolute location. *Psychological Review*, 105, 351–78.

Lee, P.U. and Tversky, B. (in preparation) Changing spatial perspective exacts a cost in reading times.

Leibowitz, H.W., Guzy, L.T., Peterson, E. and Blake, P.T. (1993) Quantitative perceptual estimates: Verbal versus nonverbal retrieval techniques. *Perception*, 22, 1051–60.

Levelt, W.J.M. (1982a) Cognitive styles in the use of spatial direction terms. In Jarvella, R.J. and Klein, W. (eds), *Speech, Place, and Action*. Chichester: Wiley, pp. 251–68.

—— (1982b) Linearization in describing spatial networks. In Peters, S. and Saarinen, E. (eds), *Processes, Beliefs, and Questions*. Dordrecht: Reidel, pp. 199–220.

—— (1984) Some perceptual limitations on talking about space. In van Doorn, A.J., van der Grind, W.A. and Koenderink, J.J. (eds), *Limits on Perception*. Utrecht: VNU Science, pp. 323–58.

—— (1989) *Speaking: From Intention to Articulation*. Cambridge, MA: MIT.

Levinson, S. (1996) Frames of reference and Molyneux's question: Cross-linguistic evidence. In Bloom, P., Peterson, M.A., Nadel, L. and Garrett, M. (eds), *Space and Language*. Cambridge, MA: MIT, pp. 109–69.

Linde, C. and Labov, W. (1975) Spatial structures as a site for the study of language and thought. *Language*, 51, 924–39.

Lloyd, R. (1989) Cognitive maps: Encoding and decoding information. *Annals of the Association of American Geographers*, 79, 101–24.

—— and Heivly, C. (1987) Systematic distortions in urban cognitive maps. *Annals of the Association of American Geographers*, 77, 191–207.

Loftus, E.F. and Palmer, J.C. (1974) Reconstruction of automobile destruction: An example of the interaction between language and memory. *Journal of Verbal Learning and Verbal Behavior*, 13, 585–9.

Lynch, K. (1960) *The Image of the City*. Cambridge, MA: MIT.

Mainwaring, S.D., Tversky, B., Ogishi, M. and Schiano, D.J. (1999) Descriptions of simple spatial scenes in English and Japanese. Manuscript submitted for publication.

Maki, R.H. (1981) Categorization and distance effects with spatial linear orders. *Journal of Experimental Psychology: Human Learning and Memory*, 7, 15–32.

Mandler, J.M., Seegmiller, D. and Day, J. (1977) On the coding of spatial information. *Memory and Cognition*, 5, 10–16.

Mark, D.M. (1992) Spatial metaphors for human-computer interaction. In Bresnahan, P., Corwin, E. and Cowen, D. (eds), *Proceedings of the Fifth International Symposium on Spatial Data Handling*. Charleston, SC: pp. 104–12.

—— Freksa, C., Hirtle, S.C., Lloyd, R. and Tversky, B. (in press) Cognitive models of geographic space. *International Journal of Geographical Information Science*.

McNamara, T.P. and Diwadkar, V.A. (1997) Symmetry and asymmetry of human spatial memory. *Cognitive Psychology*, 34, 160–90.

Milgram, S. and Jodelet, D. (1976) Psychological maps of Paris. In Proshansky, H., Ittelson, W. and Rivlin, L. (eds), *Environmental Psychology*, 2nd edn. New York: Holt, Rinehart and Winston, pp. 104–12.

Miller, G.A. and Johnson-Laird, P.N. (1976) *Language and Perception*. Cambridge, MA: Harvard.

Moar, I. and Bower, G.H. (1983) Inconsistency in spatial knowledge. *Memory and Cognition*, 11, 107–13.

Montello, D.R. (1993) Scale and multiple psychologies of space. In Frank, A.U. and Campari, I. (eds), *Spatial Information Theory: A Theoretical Basis for GIS*. Berlin: Springer-Verlag, pp. 312–21.

Morrison, J.B. and Tversky, B. (1997) Body schemas. *Proceedings of the Meetings of the Cognitive Science Society*. Mahwah, NJ: Erlbaum, pp. 525–9.

Newcombe, N. and Liben, L. (1982) Barrier effects in the cognitive maps of children and adults. *Journal of Experimental Child Psychology*, 34, 46–58.

Poulton, E.C. (1989) *Bias in Quantifying Judgments*. Hillsdale, NJ: Erlbaum.

Presson, C.C. and Montello, D.R. (1994) Updating after rotational and translational body movements: Coordinate structure of perspective space. *Perception*, 23, 1447–55.

Reed, C.L. and Farah, M.J. (1995) The psychological reality of the body schema: A test with normal participants. *Journal of Experimental Psychology: Human Perception and Performance*, 21, 334–43.

Rieser, J.J. (1989) Access to knowledge of spatial structure at novel points of observation. *Journal of Experimental Psychology: Learning, Memory and Cognition*, 15, 1157–65.

Rosch, E. (1975) Cognitive reference points. *Cognitive Psychology*, 7, 532–47.

Sadalla, E.K., Burroughs, W.J. and Staplin, L.J. (1980) Reference points in spatial cognition. *Journal of Experimental Psychology: Human Learning and Memory*, 5, 516–28.

—— and Magel, S.G. (1980) The perception of traversed distance. *Environment and Behavior*, 12, 65–79.

—— and Staplin, L.J. (1980a) An information storage model for distance cognition. *Environment and Behavior*, 12, 183–93.

—— and Staplin, L.J. (1980b) The perception of traversed distance: Intersections. *Environment and Behavior*, 12, 167–82.

Schooler, J.W. and Engstler-Schooler, T.Y. (1991) Verbal overshadowing of visual memories: Some things are better left unsaid. *Cognitive Psychology*, 22, 36–71.

Shanon, B. (1983) Answers to where-questions. *Discourse Processes*, 6, 319–52.

Shepard, R.N. (1967) Recognition memory for words, sentences and pictures. *Journal of Verbal Learning and Verbal Behavior*, 6, 156–63.

Siegel, A.W. and White, S.H. (1975) The development of spatial representations of large-scale environments. In Reese, H.W. (ed.), *Advances in Child Development and Behavior*, 10, New York: Academic Press, pp. 9–55.

Simons, D.J. (1996) In sight, out of mind: When object representations fail. *Psychological Science*, 7, 301–5.

Standing, L. (1973) Learning 10,000 pictures. *Quarterly Journal of Experimental Psychology*, 25, 207–22.

Stevens, A. and Coupe, P. (1978) Distortions in judged spatial relations. *Cognitive Psychology*, 10, 422–37.

Streeter, L.A., Vitello, D. and Wonsiewicz, S.A. (1985) How to tell people where to go: Comparing navigational aids. *International Journal of Man-Machine Studies*, 22, 549–62.

Takei, Y., Grasso, R., Amorim, M.-A. and Berthoz, A. (1997) Circular trajectory formation during blind locomotion: A test for path integration and motor memory. *Experimental Brain Research*, 115, 361–8.

—— Grasso, R. and Berthoz, A. (1996) Quantitative analysis of human walking trajectory on a circular path in darkness. *Brain Research Bulletin*, 40, 491–6.

Talmy, L. (1983) How language structures space. In Pick, H. and Acredolo, L. (eds), *Spatial Orientation: Theory, Research and Application*. New York: Plenum, pp. 225–82.

—— (1988) The relation of grammar to cognition. In Rudzka-Ostyn, B. (ed.), *Topics in Cognitive Linguistics*. Philadelphia, PA: John Benjamins, pp. 165–207.

Tappe, H. and Habel, C. (1998) Verbalization of dynamic sketch maps: Layers of representation and their interaction. Manuscript.

Taylor, H.A., Naylor, S.J. and Chechile, N.A. (1999) Goal-specific influences on the representation of spatial perspectives. *Memory and Cognition*, 27, 309–19.

—— and Tversky, B. (1992a) Descriptions and depictions of environments. *Memory and Cognition*, 20, 483–96.

—— and —— (1992b) Spatial mental models derived from survey and route descriptions. *Journal of Memory and Language*, 31, 261–82.

—— and —— (1996) Perspective in spatial descriptions. *Journal of Memory and Language*, 35, 371–91.

Thorndyke, P. (1981) Distance estimation from cognitive maps. *Cognitive Psychology*, 13, 526–50.

—— and Hayes-Roth, B. (1982) Differences in spatial knowledge acquired from maps and navigation. *Cognitive Psychology*, 14, 560–89.

Tolman, E.C. (1948) Cognitive maps in rats and men. *Psychological Review*, 55, 189–208.

Tversky, A. and Gati, I. (1978) Studies of similarity. In Rosch, E. and Lloyd, B.B. (eds), *Cognitive and Categorization*. Hillsdale, NJ: Erlbaum, pp. 79–98.

Tversky, B. (1981) Distortions in memory for maps. *Cognitive Psychology*, 13, 407–33.

—— (1991) Spatial mental models. In Bower, G.H. (ed.), *The Psychology of Learning and Motivation: Advances in Research and Theory*, vol. 27. New York: Academic Press, pp. 109–45.

—— (1992) Distortions in cognitive maps. *Geoforum*, 23, 131–8.

—— (1993) Cognitive maps, cognitive collages, and spatial mental models. In Frank, A.U. and Campari, I. (eds), *Spatial Information Theory: A Theoretical Basis for GIS*. Berlin: Springer-Verlag, pp. 14–24.

—— (1996) Spatial constructions. In Stein, N., Ornstein, P., Tversky, B. and Brainerd, C. (eds), *Memory for Emotion and Everyday Events*. Mahwah, NJ: Erlbaum, pp. 181–208.

—— (in press) Spatial memory. In Tulving, E. and Craik, F.I.M. (eds), *Handbook of Memory*. New York: Oxford University Press.

—— and Hemenway, K. (1983) Categories of scenes. *Cognitive Psychology*, 15, 121–49.

—— and Lee, P.U. (1998) How space structures language. In Freksa, C., Habel, C. and Wender, K.F. (eds), *Spatial Cognition: An Interdisciplinary Approach to Representation and Processing of Spatial Knowledge*. Berlin: Springer-Verlag, pp. 157–75.

—— and Marsh, E. (in press) Biased retellings of events yield biased memories. *Cognitive Psychology*.

—— Morrison, J., Franklin, N. and Bryant, D.J. (in press) Three spaces of spatial cognition. *Professional Geographer*.

—— and Schiano, D. (1989) Perceptual and conceptual factors in distortions in memory for maps and graphs. *Journal of Experimental Psychology: General*, 118, 387–98.

—— Taylor, H.A. and Mainwaring, S. (1997) Langage et perspective spatiale. (Language and spatial perspective) In Denis, M. (ed.), *Langage et Cognition Spatiale*, Paris: Masson, pp. 25–49.

Ungerleider, L.G. and Mishkin, M. (1982) Two cortical visual systems. In Ingle, D.J., Mansfield, R.J.W. and Goodale, M.A. (eds), *The Analysis of Visual Behavior*. Cambridge, MA: MIT, pp. 549–86.

Ullmer-Ehrich, V. (1982) The structure of living space descriptions. In Jarvella, R.J. and Klein, W. (eds), *Speech, Place and Action*. New York: Wiley, pp. 219–49.

Wilton, R.N. (1979) Knowledge of spatial relations: The specification of information used in making inferences. *Quarterly Journal of Experimental Psychology*, 31, 133–46.

Wright, P., Lickorish, A., Hull, A. and Ummelen, N. (1995) Graphics in written directions: Appreciated by readers but not by writers. *Applied Cognitive Psychology*, 9, 41–59.

Wunderlich, D. and Reinelt, R. (1982) How to get there from here. In Jarvella, R.J. and Klein, W. (eds), *Speech, Place and Action*. New York: Wiley, pp. 183–201.

Zacks, J. and Tversky, B. (1999) Event structure. Manuscript submitted for publication.

4 Cognitive mapping and spatial decision-making

Tommy Gärling and Reginald G. Golledge

Introduction

By moving from one place to another people find means of satisfying their needs. Previous research (Evans, 1980; Gärling and Golledge, 1989; Golledge, 1987) has investigated how such transitions are made. Acquired knowledge of the environment represented in memory appears to play an important role by making it possible for people to form and execute travel plans (Gärling, 1995; Gärling *et al.*, 1984; Russell and Ward, 1982).

Our focus in this chapter is spatial decision-making and, within this broader topic, the choices that people frequently make of where to move. What factors determine these choices? How are they made? These are the questions we raise and examine in the next section.

All decision-making involves choices of future courses of action. Therefore, these courses of action as well as their outcomes need to be represented in thought and imagination. In the second section we analyse the role knowledge of the environment (or cognitive maps) play in spatial decision-making. In a third section we review related previous research. In the final section we point out what we believe are promising lines of future research.

Spatial decision-making

Different approaches to the study of decision-making are labelled *normative*, *prescriptive*, and *descriptive* (Kleindorfer *et al.*, 1993). The normative approach exemplified by economic choice theory (Keeney and Raiffa, 1976; von Neumann and Morgernstern, 1944) aims at defining optimal decisions. The goal of prescriptive approaches is to advise human decision-makers about how to make optimal decisions given that they possess a limited cognitive capacity to do so. Descriptive approaches aim at illuminating how decisions are actually made. In all these approaches to decision-making, a choice is conceptualized as if it is one of several alternatives whose outcomes are possible to specify as sets of attributes. In expected utility and multi-attribute utility theories, it is assumed that a utility expressed on an interval or ratio scale is assigned to each outcome (or attribute level). A choice rule specifies the best alternative as that which maximizes utility, that is, the alternative with the highest sum of utilities assigned to outcomes or attribute levels.

A basic tenet of descriptive theories is that human decision-makers are adaptive systems with a limited capacity to adapt. Because of their limited capacity, adaptation is achieved by approximate methods (Payne *et al.*, 1993; Simon, 1982, 1990). Whereas normative theories specify what is an optimal adaptation (decision strategy), descriptive theories focus on approximate (heuristic) decision strategies which people actually use. They furthermore aim at disentangling the cognitive capacity limits and affective states or reactions which are obstacles to adaptation. We are primarily concerned with descriptive theories of spatial decision-making.

An important distinction has been made between structural and process models of decision-making (Payne, 1976; Svenson, 1979). Structural models focus on how observed choices (or preference judgements) are related to the attributes of the different choice alternatives, characteristics of the decision-maker, and possibly situational factors (e.g., time pressure, involvement). In contrast, process models attempt to specify the different cognitive and affective processes resulting in the choice. Process and structural models are not incompatible (Einhorn *et al.*, 1979; Frisch and Clemen, 1994) since the former is an explication of the latter. Process models may also vary widely with respect to the degree of detail in which they describe the process. Production systems (Anderson, 1993; Newell and Simon, 1972; Payne *et al.*, 1993) consisting of sequences of instructions in the form of 'IF (condition 1, ... , condition *n*), THEN (action 1, ... , action *m*)' are frequently used to model the process of making a choice, whereas mathematical-statistical methods are frequently used in developing structural models.

In this chapter we emphasize the process of making spatial decisions. A crude representation of this process includes the stages of generating the decision alternatives by: retrieving information about the environment which is externally accessible or is accessible in a cognitive map, representing the decision alternatives in memory, evaluating the decision alternatives, applying a decision strategy or rule, and implementing the decision (Figure 4.1).

Many spatial decisions are related to movement in the environment. However, not all movement-related decisions are spatial. In order to qualify as spatial decision-making, direction or location is necessarily an attribute of the decision alternatives. This does not exclude non-spatial attributes. Limiting ourselves to goal-directed movement, spatial decisions form a hierarchy starting at the highest level with the choice of a place where to move. This choice is then implemented leading to a choice of a path (linked routes) to the chosen place, and finally fine-grained choices of directions when the chosen path is followed (Figure 4.2). Thus, it is implied that choices of places where to move precede choices of path which in turn precede

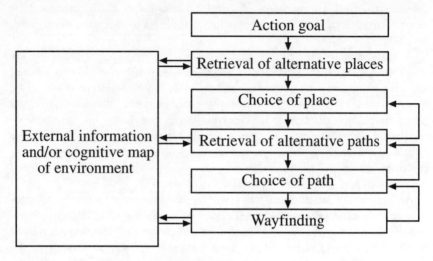

Figure 4.1 Stages of spatial decision-making.

wayfinding decisions. However, it is not invariably so since modifications may be made while the original decisions are being executed.

Both spatial and non-spatial attributes are considered in spatial decisions (Timmermans and Golledge, 1990). The former include the correlated attributes of travel distance, travel time, and energy cost of travel. In general these attributes constitute a disutility whereas non-spatial attributes either refer to the attractiveness of the place or the attractiveness of the activities which can be performed there (utility), or dangers or threats (disutility) associated with the place or activity in the place. In the former case the question is whether attractiveness of the place offsets the cost of moving to it; in the latter case whether the cost is small enough to avoid the place in favour of a safer one.

Choices of different places are frequently linked. A person who wants to move from home to A and B does not necessarily move from home to A and back, then from home to B and back. Rather, he or she moves from home to A, from A to B, and from B back home. An analysis of spatial decision-making therefore needs to be extended to multiple choices. In the general case there are *n* places from which it is possible to form combinations consisting of single places, all subsets of two places in two possible orders, all subsets of three places in six possible orders, and so forth to the set of *n* places in *n*! possible orders. Given that multiple choices are made simultaneously so that at least some combinations are compared and evaluated, this raises the question of 'how are utilities of attribute levels defined for combinations of places?' An approach to answering this question is exemplified in the work of Timmermans (1988). He assumes that combinations of alternatives have the same attributes as single alternatives and that

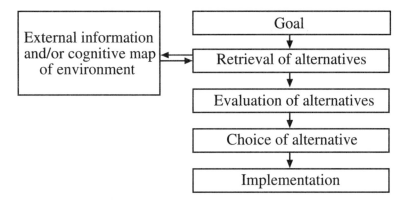

Figure 4.2 A hierarchy of spatial decisions.

utilities are related to attribute levels in the same way. However, attribute levels may combine differently. In some cases they are summed. This rule may in general apply to costs or disutilities (e.g., travel distance, time, or energy), or to utilities where two or more places are necessary to satisfy a need. Such places are complementary. In the case of utility the concept of satiation may also be emphasized. A need is possibly satisfied in one place, in which case the attractiveness or utility of other places is zero. Thus, some places are substitutable. In this case the attribute levels of combinations may be the averages of the attribute levels of single places.

A second question asks how decisions are made? Previous research (Ford *et al.*, 1989; Payne *et al.*, 1993; Svenson, 1979) has identified several different decision strategies among which a decision-maker chooses. It has been proposed that an accuracy–effort trade-off determines such choices (Payne *et al.*, 1993). However, several other factors have also been shown to affect choices of decision strategy. They include limited working memory (Newell and Simon, 1972), mood state (Isen, 1987; Loewenstein, 1996), time pressure (Svenson and Maule, 1993), and lack of knowledge (Camerer and Johnson, 1991). Choices of decision strategy are furthermore not necessarily deliberate but may constitute learned adaptations in a particular ecology (Gigerenzer and Goldstein, 1996). Two classes of decision strategies or rules contrast the normative decision rule (additive utility) to non-compensatory or heuristic decision rules (Timmermans, 1983). The latter include the satisficing (or conjunct) (Simon, 1955), elimination-by-aspects (Tversky, 1972), and lexicographic decision rules (Tversky, 1969). Decision-makers using non-compensatory decision rules avoid trade-offs which frequently are difficult to make. Another feature of non-compensatory decision rules is that they require less effort since they do not need all information to be processed (Payne *et al.*, 1988). Important determinants of the choice of non-compensatory decision strategies are the number of alternatives and the number of attributes making up the decision problem. Single spatial decisions involving a choice between a limited number of places are therefore more likely to be made by means of an additive utility or compensatory decision rule. Thus, as assumed in gravity-type models of spatial choice (Fotheringham, 1986, 1988; Rushton, 1969; Sheppard, 1978, 1980), the utility of a place may be traded off against disutility or cost of reaching it. However, when the number of alternatives increases, non-compensatory decision strategies probably play a more important role. One consequence may be that the most salient attributes receive an increased amount of attention at the expense of other attributes. For instance, minimizing negative outcomes (e.g., cost) may therefore become more salient.

A tentative process model is illustrated in Figure 4.3. Some selection of alternative places is made first. As will be discussed in the next section, a hierarchical organization of the cognitive map (Couclelis *et al.*, 1985; Golledge, 1978, 1990; Hirtle and Jonides, 1985; McNamara, 1986) facilitates decision-making by ensuring that clusters of spatially proximal locations are retrieved. In addition, a maximum distance criterion may deliberately be used to exclude places (Thill and Horowitz, 1997). In a next stage, sets of substitutable places are formed. A choice is then made between places in one of the sets if the person has a single purpose. In this choice, depending on the number of places, the person may or may not trade off different attributes. If the person must choose several places to satisfy his or her need, the first choice is followed by a second choice of a non-substitutable place, and so forth. If the number of available places is relatively small, all possible combinations may be considered. Otherwise,

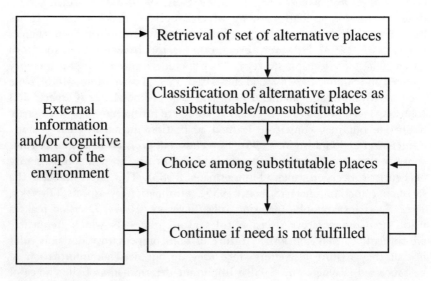

Figure 4.3 Hypothetical process of making single or multiple choices of places.

choices may be made that minimize negative outcomes (cost or distance) each time. Depending on whether trade-offs are made accurately or not, the former would lead to an optimal choice. The latter is a satisficing decision rule which does not necessarily lead to a choice which is optimal although frequently close to optimal (Payne *et al.*, 1988).

Memory representations of environments

Elements and relations

Spatial extension is a ubiquitous characteristic of human environments. At a smaller scale environments can only be perceived by successive eye fixations, at a larger scale only by moving from one place to another. The question we raise in this section is what properties of environments do people remember at different stages of experience?

A limited capacity to process and store information makes it necessary for people to be highly selective. Such selection of environmental information is already automatically accomplished at the level where patterns of physical energy impinge on the sensory organs. A deliberate selection is made at later stages when information is stored and organized in different memory systems. The end product is a memory representation or cognitive map of a particular environment. In particular, at later stages of acquisition when the cognitive map guides choices of places, choices of paths, and wayfinding decisions (Gärling *et al.*, 1984), it furthermore limits the subsequent selection of information about the environment. Therefore, the cognitive map is unlikely ever to be complete and many inaccuracies are never corrected.

Several theories have been proposed to account for how information about spatial layout is perceived, temporarily and permanently stored in memory, and later used for the performance of different tasks (e.g., Downs and Stea, 1973; Gärling *et al.*, 1984; Kitchin, 1996; Montello, 1998; Neisser, 1976). A common assumption is that the long-term memory representation becomes more complete and accurate. In addition, some theories (Hart and Moore, 1973; Siegel and White, 1975) postulate distinct, qualitatively different acquisition stages.

Incompleteness of cognitive maps resulting from deliberate selection of available information leads to a hierarchical organization of information in memory. Some theories (Hirtle and Jonides, 1985; McNamara, 1986; Stevens and Coupe, 1978) make this assumption. In the following we draw on aspects of one of them referred to as the anchor point theory (Couclelis *et al.*, 1985; Golledge, 1978, 1990).

Environmental elements represented in cognitive maps primarily constitute inanimate objects ('point' elements with limited spatial extension such as places, 'line' elements such as streets, and 'areal' elements such as districts) with some permanency in spatial location. Some of these elements, referred to as anchor points, become organizing elements of the representation by virtue of their salience. They represent the highest level in a hierarchy. It is assumed that associations are formed between anchor points in long-term memory (Anderson, 1983; Anderson and Bower, 1973). At subordinate levels elements associated with particular anchor points are only associated with each other and not with elements associated with other anchor points. Supporting empirical evidence is that anchor points function as retrieval cues providing access to other elements (Ferguson and Hegerty, 1994; Lindberg and Gärling, 1987).

The perceptual systems are tuned to picking up information about spatial layout of the environment (Gibson, 1979). It therefore seems plausible to assume that some of this information is stored in the cognitive map (Cornell *et al.*, 1994; Heft, 1996). This may be accomplished by means of forming associations between elements close in space (Hirtle and Jonides, 1985) referred to as proximity relations. Such spatially proximal clusters of places have also been shown to co-exist with places clustered according to non-spatial criteria (McNamara *et al.*, 1992). It does not, however, preclude other types of spatial relations being remembered (Gärling and Golledge, 1989; Golledge, 1987). They include crow-flight (Euclidean) distances, travel distances, and travel times (Montello, 1997). In addition, it has been shown that directions are remembered relative to points or systems of reference (Gärling *et al.*, 1986a; Sadalla *et al.*, 1980). Furthermore, level differences are remembered although possibly in distorted ways (Gärling *et al.*, 1990). Accuracy of the representation of spatial relations may vary. In general it increases with increased experience of the environment. Thus, Säisä *et al.* (1986) assumed that crow-flight distances are more accurately remembered than travel distances, and that travel distances are more accurately remembered than travel times. Although routes are traversed while learning the spatial relations between places, only the relative locations of places may initially be stored in long-term memory (Lindberg and Gärling, 1982, 1983). It is not until the level of acquisition has increased further that route distances are remembered. At this level of acquisition, travel times may be inferred from travel distances but may later develop into separate memory representations when additional experiences accrue.

The mode of acquisition has also been found to be an important determinant of how accurately and precisely different kinds of spatial relations are remembered. Anchor points, elements which are less salient, and their relations at different spatial scales are learned through direct encounters of the environment, through secondary sources such as maps, through inferences, or through combinations of these modes. For instance, learning from available maps may lead to more accurate memory of crow-flight distances and directions, whereas actual experience with the environment may lead to more accurate memory of travel distances (Evans and Pezdek, 1980; Sholl, 1987; Thorndyke and Hayes-Roth, 1982).

An element or relation between elements in the environment may be represented temporarily in working memory during acquisition. As Lindberg and Gärling (1982, 1983) found, this increases the likelihood that a spatial relation is permanently stored in memory. In connection with, for instance, wayfinding decisions, information retrieved from long-term memory may be temporarily stored in working memory. If the retrieved information is transformed, the outcome may subsequently be stored in long-term memory. For example, the number of spatial relations between elements permanently stored in memory may increase as a result of inferences of how A and C are spatially related (based on information about how A is spatially related to B and how B is spatially related to C). This increased consistency has been demonstrated in field studies of cognitive maps of large-scale environments. For instance, in a study comparing newcomers to a city, those who had been there between three months and a year, and those who had been there for two years or longer, Golledge (1978) showed that cognitive maps evolved from highly distorted non-metric representations to representations closely reflecting the real world. Gale *et al.* (1990) further explored the role of familiarity, differentiating between general familiarity and spatial familiarity. Frequency and length of exposure to the environment were found to correlate with how accurate recall was.

A cognitive map consists of declarative knowledge. Applying declarative knowledge in spatial decisions leads to procedural knowledge (choices of actions) which may co-exist with or replace the declarative knowledge. Some procedures may also be acquired independently of declarative knowledge. In general however, following Anderson (1993) it may be assumed that with practice declarative knowledge acquisition is followed by a 'compilation' stage in which the declarative knowledge is successively transformed to procedures. Thus, choices may become automatized or 'script-based' (Svenson, 1998; Verplanken *et al.*, 1997). The consequence is that procedures entailing choices are executed without much deliberation.

Decision alternatives

In general, benefits of moving to a place are traded off against the cost (distance) of moving there. At other times cost is the single choice criterion. Accurately assessing the cost requires information about the spatial layout of the environment. We offer here a set of hypotheses about constraints on accurate evaluations imposed by the properties of cognitive maps when information is accessed from such cognitive maps. In connection with making decisions about where to move, it is assumed that people construct mental models or temporary representations in working memory based on retrieval from long-term memory (Johnson-Laird, 1983; Tversky, 1991). Therefore, the kinds of mental models people construct affect their spatial decisions. For instance, under certain circumstances people are able to form analogue spatial representations (spatial images) that make possible simultaneous comparisons of several locations in space (Gärling, 1989; Gärling *et al.*, 1986b). However, constraints on this ability are imposed by the limited capacity of working memory, the nature of, and how, the information is stored in long-term memory.

In the following we first distinguish between a low, an intermediate, and a high level of acquisition of the cognitive map (Table 4.1). Derived from assumptions about how much information needs to be stored, typical acquisition mode, and load imposed on information processing in conjunction with acquisition (Lindberg and Gärling, 1982), we hypothesize that knowledge of spatial relations stored in cognitive maps develops in a certain sequence. At a low level of acquisition spatial relations between anchor points as well as between anchor points, and some other elements, consist of crow-flight distances and directions. At an intermediate level of acquisition, knowledge of travel distances is added. Finally, at a high level of acquisition, knowledge of travel times is added. We further assume that there is an increase in the number of anchor points and other elements related to anchor points as well as an increase of the accuracy and precision of their representation from the low to the intermediate acquisition stage. We then distinguish between single and multiple choices. Multiple choices are either sequential – that is, made as if they were independent single choices, in which case comparisons are limited each time to choice sets consisting of single places – or simultaneous in which case combinations of single places are compared and evaluated. Below follows a brief description of the characteristics of different cases obtained by crossing level of acquisition with single versus multiple choices:

Case IA (single spatial choices among alternative places at a low level of acquisition): Choices between places are based on crow-flight distances. The main sources of inaccuracy are that the choice set is limited, that not all locations are spatially related, and that knowledge of crow-flight distances is frequently incomplete.

Case IB (single spatial choices among alternative places at an intermediate level of acquisition): The primary differences, compared to the former case, are that more anchor points and related locations are stored, that the accuracy of the knowledge is higher, and that travel distances are known.

Case IC (single spatial choices among alternative places at a high level of acquisition): This case differs from the former in that knowledge of travel times is available. Accuracy is further increased.

Case IIA (multiple spatial choices among alternative places at a low level of acquisition): For the limited choice set of anchor points, multiple choices are possible to make on the basis of crow-flight distances. In other cases only sequences of single choices (from A to B and back, then from A to C, and so forth) are possible to make, again based on crow-flight distances. The limited capacity of working memory imposes

Table 4.1 Hypothetical cases of choices of places depending on level of acquisition of a cognitive map and whether choices are single or multiple

	Level of acquisition		
	Low	Intermediate	High
Single choice	Case IA	Case IB	Case IC
Multiple choices	Case IIA	Case IIB	Case IIC

constraints on the ability to make accurate simultaneous choices when the choice set increases.

Case IIB (multiple spatial choices among alternative places at an intermediate level of acquisition): The set of anchor points is larger, thus increasing the possibility of making simultaneous multiple choices. These choices are also more accurate. However, working memory capacity sets an upper limit. Choices of places are based on travel distances.

Case IIC (multiple spatial choices among alternative places at a high level of acquisition): This case differs from the preceding one in that travel times is the choice criterion.

The preceding analysis leads to the qualitative prediction that choices of different places in the environment are more frequently linked as the level of acquisition of cognitive maps increases. It is also predicted that choices will be more efficient since more accurate assessments of utilities to visit and costs to move to places and combinations of places become possible. Yet, even though knowledge of the environment increases, there will be an upper limit to using this knowledge set by the capacity of working memory. Thus, asymptotic degrees of chaining of movements and efficiency are expected. Further detailed predictions are only possible based on valid and reliable measures of the cognitive maps acquired of the environment.

Review of previous research[1]

As Lloyd noted in 1976, at that time less research had been devoted to the study of the links between cognitive maps, spatial preferences, and spatial choices than to the study of each of these separately. Most of the available scarce research had focused on shopping behaviour. A subsequent review (Halperin, 1988) suggested that there were few changes to this state of affairs a decade later. Not even the obvious need of assuming that the forming of a choice set of places many times depends on knowledge stored in a cognitive map appears to have been recognized (Thill, 1992).

Gärling *et al.* (1984), Russell and Snodgrass (1987), and Snodgrass *et al.* (1988) are early examples of more general attempts to conceptualize and investigate how cognitive maps affect spatial choices. In the same vein, a few studies have focused on how people solve the travelling salesperson problem which requires choosing a path to travel between a set of places which minimizes total distance (Gärling, 1999; MacGregor and Ormerod, 1996). The results indicated that pedestrians and other travellers locally minimize distances between places but do not necessarily minimize total distances. People may however sometimes do better than that if they have access to or are able to construct a spatial representation in working memory (Gärling, 1989; Gärling *et al.*, 1986b; Gärling and Gärling, 1988; Hirtle and Gärling, 1992).

A broader approach to shopping behaviour was pioneered by Downs (1970) who used methods derived from Kelly (1955) to study consumers' images of shopping centres. Subsequent research includes, among others, Crewe and Lowe (1995). In a similar way, Cadwallader (1975), Huff (1964), and Rushton (1971) discussed how perceived attractiveness of (shopping) places is traded off against the inhibiting friction of travel distance. However, many other early studies of shopping behaviour (e.g., Briggs, 1969, 1972; Day, 1976; Pacione 1976; Potter 1976a, 1976b) primarily focused on the role of spatial information stored in cognitive maps, in particular the role of cognitive distance. For instance, Thompson (1963, 1966) incorporated subjective distance into conventional gravity types of models used to explain store choice. A similar tactic was adopted by Cadwallader (1975), Burnett (1978), and MacKay *et al.* (1975) concluded that both cognitive (crow-flight) distances and travel times were important determinants of choices of shopping locations. The concept of subjective or cognitive distance appeared sufficiently significant to warrant independent research into its nature (e.g., Baird *et al.*, 1979; Burroughs and Sadalla, 1979; Golledge *et al.*, 1969; Lowrey, 1970; Sadalla and Staplin, 1980a, 1980b). Although there are exceptions, this research has shown that cognitive distance is not proportional but is a negatively accelerated power function of objective distance (Montello, 1997). Thus, short distances are overestimated and long distances are underestimated. However, whether or not this is an artificial finding has been debated (Baird *et al.*, 1979, 1982; Cadwallader, 1979; Day, 1976). Montello (1991) provides a useful summary of relevant methodological research.

The degree of curvature of the relationship between cognitive and objective distance has been shown to be affected by numerous factors. A basic finding is that increasing familiarity with the environment decreases the curvature (e.g., Golledge, 1978). Some other results such as the reported effects of direction to 'downtown' (Golledge *et al.*, 1969; Lee, 1962) may possibly be accounted for by familiarity differences. Yet, Sadalla and Staplin (1980a, 1980b) were able to show in controlled experiments that a landmark anchoring the end of a distance resulted in differences in estimates of distances towards the landmark as compared to estimates of distances from the landmark. Not many studies appear to have contrasted people to whom the environment is familiar with those to whom the environment is unfamiliar. An exception is a study by Lotan (1997) comparing simulated route decisions by different groups of drivers. The results showed that familiarity (degree of exposure to the environment) decreased variability in choices and increased skill in making detours.

There have also been speculations about the non-Euclidean nature of cognitive maps (Golledge and Hubert, 1982). Baird *et al.* (1982) noted that the often observed asymmetry of cognitive distance as well as its non-linear relation to objective, crow-flight distance make it impossible to construct integrated cognitive maps in conventional geometric spaces. If this is the case, one may expect intransitivity of spatial choices. Intransitivity of preferences for non-spatial choice alternatives has been demonstrated (e.g., Tversky, 1969). In a similar vein, replicating the results of an earlier study of undergraduate students (Moar and Carleton, 1982), Stern and Leiser (1988) found that professional drivers, whose route knowledge presumably is more complete, less frequently chose different routes from a place to another and back than non-professional drivers.

Research on the effect of cognitive distance on spatial choices has focused on the role played by systematic errors but largely neglected the role of unsystematic errors or fuzziness. An exception is Cadwallder (1975) who assumed that degree of familiarity increases the likelihood that a place is chosen. However, it is unclear if this holds at a disaggregate, individual level. A single study conducted by the authors (Gärling *et al.*, 1999b) addressed this question in an indirect way. Subjects made choices between walking and driving when distance was presented as an interval that delimited the possible distances. The results showed that choices to drive were based on the upper boundary of the interval suggesting the application of a worst-case scenario. Thus, in effect, choices of driving were made for much shorter distances than when there was no uncertainty about distance.

In transport research (Hanson, 1995) trip chaining resulting in multi-purpose, multi-stop journeys is a recognized phenomenon. Yet, in theories of travel choice such journeys are frequently ignored. In their review, Thill and Thomas (1987) point out that a typical assumption is that the sequence of choice alternatives that people select is generated by a stochastic process. Some exceptions are the theories proposed by Kitamura (1984) and Timmermans and van der Waerden (1992). Nevertheless, since the assumption is made that the utility maximization principle is valid, the generality of these theories is limited. In Borgers and Timmermans (1986) pedestrians were assumed to make choices from all possible sequences of alternative stores in a shopping mall. Existing knowledge about cognitive capacity limits indicates that this is an unrealistic assumption. In contrast, Gärling and Gärling (1988) and Gärling *et al.* (1997a, 1997b) offered alternative assumptions which are behaviourally realistic and which have received empirical support in controlled experiments. However, almost no research appears to have examined how properties of cognitive maps affect sequential choices.

Adler and Ben-Akiva (1979) assumed that choices are made between sequences of stops. By obtaining the choice set from actually observed sequences, they were able to account for differences in spatial knowledge to some extent. Still, a better method would be to use data on recall of locations. A very different approach is represented by the studies by Gärling and collaborators (Gärling, 1989; Gärling *et al.*, 1986b; Säisä and Gärling, 1987) of how subjects solve the travelling salesperson problem. This research has examined effects of load imposed on working memory, subjects' ability to construct spatial representations in working memory, and properties of spatial representations in long-term memory.

A related observation is that people strive to achieve some degree of variation when they choose places where to move. In other words, people may have a desire to choose a different alternative on the next choice occasion, even if they are satisfied with the outcome of their previous choice. This tendency may increase with increased knowledge of the environment. Evidence supporting variety-seeking in choices of places was presented by Borgers, van der Heijden, and Timmermans (1989). However, no research has so far related this tendency to the properties of people's cognitive maps.

A future research agenda

Earlier in this chapter, hypotheses were offered concerning how properties of the cognitive map of an environment affect choices about moving to alternative places. Although previous research has focused on some of the relevant issues, there are apparently only a few studies of the direct effects on spatial decision-making of level of acquisition of cognitive maps. Such studies may preferably employ an experimental approach in which acquisition level is systematically varied. Given practical and ethical constraints, in such experiments it would only be feasible to investigate small-scale environments. A surrogate method is therefore to stratify residents of a city on time in residence or some other proxy of level of acquisition.

In Gärling *et al.* (1999a) a small sample of households was asked to provide detailed information about their choices of daily activities during a week. Information was also obtained on recall of shopping locations. Time in residence was shown to account for a sizeable amount of variance in recall. However, the relationship tended to be inverted U-shaped rather than positively linear. Forgetting may possibly account for this deviation from linearity. It is plausible to assume that over time households first add more alternatives to their choice sets of shopping locations, then exclude alternatives when they find them unattractive. Alternatives which are excluded may be forgotten. Yet, in particular households who seek variety and/or make many multi-purpose, multi-stop trips may maintain a larger choice set. In a study by Biel (1973) of households who moved from one side of the town to another, many places that were usually chosen when living at the original location were still chosen for the first two weeks.

Grocery stores near the new home were the first to be substituted for previously patronized sites. Other convenience shopping stores were soon also substituted, including bank branches, post offices, dry cleaners, and others. But many people retained their previously visited sites for professional services (e.g., doctors and dentists). Thus, some patronage patterns were changed, but others were not, even after a considerable time lapse. In summary, this appears to demonstrate that the interplay of acquisition, forgetting, and maintenance of information in the cognitive map is an important, largely neglected focus of research. Clearly, spatial decisions and choices play an important role for this interplay.

It would not be possible in future research to ignore possibly large individual differences in choice sets, and how these affect spatial choices. Individual differences may largely reflect differences in interest. For instance, in many households wives are likely to be primarily responsible for grocery shopping and may therefore have larger choice sets of shopping locations. However, as discussed by Self and Golledge in Chapter 12 of this volume, men and women may also differ in how their decisions relate to the available information about the decision alternatives. For example, Montello *et al.* (1999) found that women recall more detailed cue information when learning layouts than do men, and that women tend to recognize and recall more landmarks than do men. This might imply that, for activities such as shopping, women may recall and consider more choice alternatives than do men.

Research also needs to go beyond measuring the size of choice sets. Are people more likely to make simultaneous choices of places when they have more complete and accurate spatial knowledge stored in their cognitive maps? Are their choices more accurate? Since a limited-capacity working memory imposes a constraint on information processing, these questions do not have self-evident answers. Similarly, there is no general answer to the question of how uncertainty of spatial knowledge or fuzziness affects spatial decisions. As already alluded to, expected utility theory may not supply the correct answer. Here are several research questions which have not yet been addressed.

Some research has investigated implementation of choices of places, in particular wayfinding decisions (Figure 4.2). Chown *et al.* (1995), Oatley (1977), and Warren (1994) suggested that wayfinding is a dynamic decision-making process which recurrently attempts to answer the question 'Where am I?'. The answer appears to require knowledge of spatial relations between elements in the environment as well as memory for scenes, perspectives, and scene sequences, all of which are likely to be stored in the cognitive map. However, Gärling *et al.* (1984) and Passini (1980) argued that wayfinding decisions may be conceptualized as problem solving in which incomplete information in cognitive maps is supplemented with information acquired while travelling. In a similar vein, Lawton (1996) suggested that people with insufficient memory of the environment use generalized knowledge of buildings or other elements to successfully find their way to specific places. The general implication is that acquisition of a cognitive map, making choices of places where to move, and implementing these choices is not always a linear process. Further research is needed on the interrelationship of these component processes.

Conclusions

In this chapter we have focused on how properties of cognitive maps affect spatial decision-making. From our selective review of research on spatial choice we conclude that, after the promising beginning in the seventies, little attention appears to have been paid to cognitive maps. Although there are many possible reasons for this, we believe one is that subsequent research has tended to ignore the *process* of making spatial decisions. Motivated by practical applications, the research has emphasized 'objective' predictors of spatial choices at the expense of emphasizing the details of the process. However, there is no conflict here. Therefore, we hope to see a future upsurge of interest in the process-oriented approach to the modelling of spatial choice. The need for this was actually pointed out some time ago (Thill and Thomas, 1987). Modelling the process of making spatial choice cannot ignore choice set characteristics (Thill, 1992). We hope that, in this chapter, we have instigated an impetus for drawing on cognitive mapping research in developing models of choice set formation. We have also provided some more specific ideas of how this can be done. Obviously, these ideas in themselves motivate further research and we have pointed out what we believe are valuable directions of this research.

Acknowledgements

In part this chapter was written while Tommy Gärling visited the University of California, Santa Barbara, in the winter of 1998. His visit was made possible in part by NSF grant #SBR-9514907, by funding from NCGIA, and by grant #93–315–22 from the Swedish Transport and Communications Research Board.

Notes

1   The review excludes spatial decisions leading to more permanent changes of location such as choices of migration destinations. In general it is plausible to assume that to a larger extent people search external information rather than retrieve information from their cognitive maps when making such decisions.

References

Adler, T. and Ben-Akiva, M. (1979) A theoretical and empirical model of trip chaining behavior. *Transportation Research B*, 13, 243–57.

Anderson, J.R. (1983) *The Architecture of Cognition*. Cambridge, MA: Harvard.
—— (1993) *Rules of the Mind*. Hillsdale, NJ: Erlbaum.
—— and Bower, G.H. (1973) *Associative Memory*. Washington, DC: Winston.
Baird, J.C., Merril, A.A. and Tannenbaum, J. (1979) Studies of cognitive representations of spatial relations: II. A familiar environment. *Journal of Experimental Psychology: General*, 108, 92–8.
—— Wagner, M. and Noma, E. (1982) Impossible cognitive spaces. *Geographical Analysis*, 14, 204–16.
Biel, A. (1973) Spatial aspects of extinction process: A preliminary investigation. Discussion Paper #35 (pp. 14–41), Department of Geography, Ohio State University.
Borgers, A.W.J. and Timmermans, H.J.P. (1986) A model of pedestrian route choice and demand for retail facilities within inner-city shopping areas. *Geographical Analysis*, 18, 115–23.
—— van der Heijden, R.E.C.M. and Timmermans, H.J.P. (1989) A variety seeking model of spatial choice-behaviour. *Environment and Planning A*, 21, 1037–48.
Briggs, R. (1969) Scaling of preferences for spatial location: An example using shopping centers. Unpublished MA thesis, Columbus, OH: Ohio State University.
Briggs, R. (1972) Cognitive distance in urban space. Unpublished PhD dissertation. Columbus, OH: Ohio State University.
Burnett, P. (1978) *Choice and Constraints Oriented Modeling: Alternative Approaches to Travel Behavior*. Evanston, IL: Department of Geography, Northwestern University.
Burroughs, W. and Sadalla, E. (1979) Asymmetries in distance cognition. *Geographical Analysis*, 11, 414–21.
Cadwallader, M.T. (1975) A behavioral model of consumer spatial decision making. *Economic Geography*, 51, 339–49.
—— (1979) Problems in cognitive distance: Implications for cognitive mapping. *Environment and Behavior*, 11, 559–76.
Camerer, C.F. and Johnson, E.J. (1991) The process-performance paradox in expert judgment. In Ericson, K.A. and Smith, J. (eds), *Toward a General Theory of Expertise*. Cambridge: Cambridge University Press, pp. 195–217.
Chown, E., Kaplan, S. and Kortenkamp, D. (1995) Prototypes, Location, and Associative Networks (PLAN): Towards a unified theory of cognitive mapping. *Cognitive Science*, 19, 1–51.
Cornell, E., Heth, D. and Alberts, D.M. (1994) Place recognition and wayfinding by children and adults. *Memory and Cognition*, 22, 633–43.
Couclelis, H., Golledge, R.G., Gale, N. and Tobler, W. (1985) Exploring the anchor-point hypothesis of spatial cognition. *Journal of Environmental Psychology*, 5, 99–122.
Crewe, L. and Lowe, M. (1995) Gap on the map? Towards a geography of consumption and identity. *Environment and Planning A*, 27, 1877–98.
Day, R.A. (1976) Urban distance cognitions: Review and contribution. *The Australian Geographer*, 13, 193–200.
Downs, R.M. (1970) The cognitive structure of an urban shopping center. *Environment and Behavior*, 2, 13–39.
—— and Stea, D. (1973) Theory. In Downs, R.M. and Stea, D. (eds), *Image and Environment*. London: Arnold, pp. 1–7.
Einhorn, H.J., Kleinmuntz, D.N. and Kleinmuntz, B. (1979) Linear regression and process-tracing models of judgment. *Psychological Review*, 86, 465–85.

Evans, G.W. (1980) Environmental cognition. *Psychological Bulletin*, 88, 259–87.

—— and Pezdek, K. (1980) Cognitive mapping: Knowledge of real-world distance and location information. *Journal of Experimental Psychology: Human Learning and Memory*, 6, 13–24.

Ferguson, E.L. and Hegerty, M. (1994) Properties of cognitive maps constructed from text. *Memory and Cognition*, 22, 455–73.

Ford, J.K., Schmitt, N., Schechtman, S.L., Hults, B.M. and Doherty, M.L. (1989) Process tracing methods: Contributions, problems, and neglected research questions. *Organizational Behavior and Human Decision Processes*, 43, 75–117.

Fotheringham, A.S. (1986) Modelling hierarchical destination choice. *Environment and Planning A*, 18, 401–18.

—— (1988) Consumer store choice and choice set definition. *Marketing Science*, 7, 299–310.

Frisch, D. and Clemen, R. (1994) Beyond expected utility: Rethinking behavioral decision research. *Psychological Bulletin*, 116, 46–54.

Gale, N.D., Golledge, R.G., Halperin, W.C. and Couclelis, H. (1990) Exploring spatial familiarity. *Professional Geographer*, 42: 299–313.

Gärling, T. (1989) The role of cognitive maps in spatial decisions. *Journal of Environmental Psychology*, 9, 269–78.

—— (1995) How do urban residents acquire, mentally represent, and use knowledge of spatial layout? In Gärling, T. (ed.), *Readings in Environmental Psychology: Urban Cognition*. London: Academic Press, pp. 1–12.

—— (1999) Human information processing in sequential spatial choice. In Golledge, R.G. (ed.), *Wayfinding: Cognitive Maps and Spatial Behavior*. Baltimore, MD: Johns Hopkins, pp. 81–98.

—— Boe, O. and Golledge, R.G. (1999a) *Determinants of Drivers' Destination and Mode Choices*. Manuscript.

—— Böök, A. and Lindberg, E. (1984) Cognitive mapping of large-scale environments: The interrelationship of action plans, acquisition, and orientation. *Environment and Behavior*, 16, 3–34.

—— Böök, A., Lindberg, E. and Arce, C. (1990) Is elevation encoded in cognitive maps of large-scale environments. *Journal of Environmental Psychology*, 10, 341–51.

—— Doherty, S. and Golledge, R.G. (1999b) Cognitive maps and choices of shopping locations: An analysis of household activity scheduling. Manuscript.

—— and Gärling, E. (1988) Distance minimization in downtown pedestrian shopping behavior. *Environment and Planning A*, 20, 547–54.

—— Gillholm, R., Romanus, J. and Selart, M. (1997a) Interdependent activity and travel choices: Behavioral principles of integration of choice outcomes. In Ettema, D. and Timmermans, H.P.J. (ed.), *Activity-Based Approaches to Travel Analysis*. Oxford: Pergamon, pp. 135–50.

—— and Golledge, R.G. (1989) Environmental perception and cognition. In Zube, E.H. and Moore, G.T. (eds), *Advances in Environment, Behavior, and Design*, vol. 2, New York: Plenum, pp. 203–36.

—— Karlsson, N., Romanus, J. and Selart, M. (1997b) Influences of the past on choices of the future. In Ranyard, R., Crozier, R. and Svenson, O. (eds), *Decision Making: Cognitive Models and Explanations*. London: Routledge, pp. 167–88.

—— Lindberg, E., Carreiras, M. and Böök, A. (1986a) Reference systems in cognitive maps. *Journal of Environmental Psychology*, 6, 1–18.

—— Säisä, J., Böök, A., and Lindberg, E. (1986b) The spatiotemporal sequencing of everyday activities in the large-scale environment. *Journal of Environmental Psychology*, 6, 261–80.

Gibson, J.J. (1979) *The Ecological Approach to Visual Perception.* Boston, MA: Houghton-Mifflin.

Gigerenzer, G. and Goldstein, D.G. (1996) Reasoning the fast and frugal way: Models of bounded rationality. *Psychological Review*, 98, 506–28.

Golledge, R.G. (1978) Learning about urban environments. In Carlstein, T., Parkes, D. and Thrift, N. (eds), *Timing Space and Spacing Time*, vol. 1. London: Arnold, pp. 76–89.

—— (1987) Environmental cognition. In Stokols, D. and Altman, I. (eds), *Handbook of Environmental Psychology*, vol. 1. New York: Wiley, pp. 131–74.

—— (1990) The conceptual and empirical basis of a general theory of spatial knowledge. In Fischer, M., Nijkamp P. and Papageorgiou, Y. (eds), *Spatial Choices and Processes*. Amsterdam: North-Holland, pp. 147–68.

—— Briggs, R. and Demko, D. (1969) The configuration of distances in intra-urban space. *Proceedings of the Association of American Geographers*, 1, 60–5.

—— and Hubert, L.J. (1982) Some comments on non-euclidean mental maps. *Environment and Planning A*, 14, 107–18.

Halperin, W.C. (1988) Current topics in behavioral modelling of consumer choice. In Golledge, R.G. and Timmermans, H.J.P. (eds), *Behavioral Modelling in Geography and Planning*. London: Croom Helm, pp. 1–26.

Hanson, S. (1995) Getting there: Urban transportation in context. In Hanson, S. (ed.), *The Geography of Urban Transportation*, 2nd edn, New York: Guilford, pp. 3–25.

Hart, R.A. and Moore, G.T. (1973) The development of spatial cognition: A review. In Downs, R.M. and Stea, D. (eds), *Image and Environment*. London: Arnold, pp. 246–88.

Heft, H. (1996) The ecological approach to navigation: A Gibsonian perspective. In Portugali, J. (ed.), *The Construction of Cognitive Maps*. Dordrecht: Kluwer, pp. 105–32.

Hirtle, S.C. and Gärling, T. (1992) Heuristic rules for sequential spatial decisions. *Geoforum*, 23, 227–38.

—— and Jonides, J. (1985) Evidence of hierarchies in cognitive maps. *Memory and Cognition*, 13, 208–17.

Huff, D.L. (1964) Defining and estimating a trade area. *Journal of Marketing*, 28, 34–8.

Isen, A.M. (1987) Positive affect, cognitive processes, and social behavior. In Berkowitz, L. (ed.), *Advances in Experimental Social Psychology*, vol. 20, San Diego, CA: Academic, pp. 203–53.

Johnson-Laird, P.N. (1983) *Mental Models*. Cambridge, MA: Harvard.

Keeney, R.L. and Raiffa, H. (1976) *Decisions with Multiple Objectives: Preferences and Value Tradeoffs*. New York: Wiley.

Kelly, G.A. (1955) *The Psychology of Personal Constructs*. New York: Norton.

Kitamura, R. (1984) Incorporating trip chaining in analysis of destination choice. *Transportation Research B*, 18, 67–81.

Kitchin, R.M. (1996) Increasing the integrity of cognitive mapping research: Appraising conceptual schemata of environment-behavior interaction. *Progress in Human Geography*, 20, 56–84.

Kleindorffer, P.R., Kunreuther, H.C. and Schoemaker, P.J.H. (1993) *Decision Sciences: An Integrative Perspective*. New York: Cambridge University Press.

Lawton, C.A. (1996) Strategies for indoor wayfinding: The role of orientation. *Journal of Environmental Psychology*, 16, 137–45.

Lee, T.R. (1962) Brennan's law of shopping behavior. *Psychological Reports*, 2, 662.

Lindberg, E. and Gärling, T. (1982) Acquisition of locational information about reference points during locomotion: The role of central information processing. *Scandinavian Journal of Psychology*, 23, 207–18.

—— and Gärling, T. (1983) Acquisition of different types of locational information in cognitive maps: Automatic or effortful processing? *Psychological Research*, 45, 19–38.

—— and Gärling, T. (1987) Memory for spatial location in two-dimensional arrays. *Acta Psychologica*, 64, 151–66.

Lloyd, R.E. (1976) Cognition, preference, and behavior in space. *Economic Geography*, 52, 241–53.

Loewenstein, G. (1996) Out of control: Visceral influences on behavior. *Organizational Behavior and Human Decision Processes*, 65, 272–92.

Lotan, T. (1997) Effects of familiarity on route choice behavior in the presence of information. *Transportation Research C*, 5, 225–43.

Lowrey, R.A. (1970) Distance concepts of urban residents. *Environment and Behavior*, 2, 57–73.

MacGregor, J.N. and Ormerod, T. (1996) Human performance on the traveling salesman problem. *Perception and Psychophysics*, 58, 527–39.

MacKay, D.B., Olshavsky, R.W. and Sentell, G. (1975) Cognitive maps and spatial behaviors of consumers. *Geographical Analysis*, 7, 19–34.

McNamara, T.P. (1986) Mental representations of spatial relations. *Cognitive Psychology*, 18, 87–121.

—— T.P., Halpin, J. and Hardy, J. (1992) The representation and integration in memory of spatial and nonspatial information. *Memory and Cognition*, 20, 519–32.

Moar, I. and Carleton, I.R. (1982) Memory for routes. *Quarterly Journal of Experimental Psychology*, 34A, 381–94.

Montello, D.R. (1991) The measurement of cognitive distance: Methods and construct validity. *Journal of Environmental Psychology*, 11, 101–22.

—— (1997) The perception and cognition of environmental distance: Direct sources of information. In Hirtle, S.C. and Frank, A.U. (eds), *Spatial Information Theory: A Theoretical Basis for GIS. Lecture Notes in Cognitive Science*, vol. 1329. Berlin: Springer-Verlag, pp. 297–311.

—— (1998) A new framework for understanding the acquisition of spatial knowledge in large-scale environments. In Egenhofer, M.J. and Golledge, R.G. (eds), *Spatial and Temporal Reasoning in Geographic Information Systems*. Oxford: Oxford University Press, pp. 143–54.

—— Lovelace, K.L., Golledge, R.G. and Self, C.M. (1999) Sex-related differences and similarities in geographic and environmental spatial abilities. Manuscript.

Neisser, U. (1976) *Cognition and Reality*. San Francisco, CA: Freeman.

Newell, A. and Simon, H.A. (1972) *Human Problem Solving*. Englewood Cliffs, NJ: Prentice-Hall.

Oatley, K.G. (1977) Inference, navigation, and cognitive maps. In Johnson-Laird, P.N. and Wason, P.C. (eds), *Thinking: Readings in Cognitive Science.* Cambridge: Cambridge University Press.

Pacione, M. (1976) A measure of the attraction factor: A possible alternative. *Area*, 6, 279–82.

Passini, R. (1980) Wayfinding in complex buildings: An environmental analysis. *Man-Environment Systems*, 10, 31–40.

Payne, J.W. (1976) Task complexity and contingent processing in decision making: An information search and protocol analysis. *Organizational Behavior and Human Performance*, 16, 366–87.

—— Bettman, J.R. and Johnson, E.J. (1988) Adaptive strategy selection in decision making. *Journal of Experimental Psychology: Learning, Memory, and Cognition*, 14, 534–52.

—— Bettman, J.R. and Johnson, E.J. (1993) *The Adaptive Decision Maker.* New York: Cambridge University Press.

Potter, R.B. (1976a) Directional bias within the usage and perceptual fields of urban consumers. *Psychological Reports*, 38, 988–90.

—— (1976b) Spatial nature of consumer usage and perceptual fields. *Perceptual and Motor Skills*, 43, 1185–88.

Rushton, G. (1969) Analysis of spatial behavior by revealed space preference. *Annals of the Association of American Geographers*, 59, 391–400.

—— (1971) Preference and choice in different environments. *Proceedings of the Association of American Geographers*, 3, 146–50.

Russell, J.A. and Snodgrass, J. (1987) Emotions in the environment. In Stokols, D. and Altman, I. (eds), *Handbook of Environmental Psychology*, vol. 1. New York: Wiley, pp. 255–80.

—— and Ward, L.M. (1982) Environmental psychology. *Annual Review of Psychology*, 33, 651–88.

Sadalla, E.K., Burroughs, W.J. and Staplin, L.J. (1980) Reference points in spatial cognition. *Journal of Experimental Psychology: Human Learning and Memory*, 6, 516–28.

—— and Staplin, L.J. (1980a) An information storage model for distance cognition. *Environment and Behavior*, 12, 183–93.

—— and —— (1980b) The perception of traversed distance: Intersections. *Environment and Behavior*, 12, 167–82.

Säisä, J. and Gärling, T. (1987) Sequential spatial choices in the large-scale environment. *Environment and Behavior*, 19, 614–35.

—— Svensson-Gärling, A., Gärling, T. and Lindberg, E. (1986) Intraurban cognitive distance: The relations between judgments of straight-line distances, travel distances, and travel times. *Geographical Analysis*, 18, 167–74.

Sheppard, E.S. (1978) Theoretical underpinnings of the gravity hypothesis. *Geographical Analysis*, 10, 386–401.

—— (1980) Location and the demand for travel. *Geographical Analysis*, 12, 2: 111–28.

Sholl, M.J. (1987) Cognitive maps as orienting schema. *Journal of Experimental Psychology: Learning, Memory, and Cognition*, 13, 615–28.

Siegel, A.W. and White, S.H. (1975) The development of spatial representations of large-scale environments. In Reese, W.H. (ed.), *Advances in Child Development and Behavior*, vol. 10. New York: Academic, pp. 9–55.

Simon, H.A. (1955) A behavioral model of rational choice. *Quarterly Journal of Economics*, 69, 99–118.

—— (1982) *Models of Bounded Rationality. Vol. 2: Behavioral Economics and Business Organization*. Cambridge, MA: MIT.

—— (1990) Invariants of human behavior. *Annual Review of Psychology*, 41, 1–19.

Snodgrass, J., Russell, J. and Ward, L. (1988) Planning, mood, and place-liking. *Journal of Environmental Psychology*, 8, 209–22.

Stern, E. and Leisser, D. (1988) Levels of spatial knowledge and urban travel modeling. *Geographical Analysis*, 20, 140–55.

Stevens, A. and Coupe, P. (1978) Distortions in judged spatial relations. *Cognitive Psychology*, 10, 422–37.

Svenson, O. (1979) Process descriptions of decision making. *Organizational Behavior and Human Performance*, 23, 86–112.

—— (1998) The perspective from behavioral decision theory on modeling travel choice. In Gärling, T., Laitila, T. and Westin, K. (eds), *Theoretical Foundations of Travel Choice Modeling*. Amsterdam: Elsevier, pp. 141–72.

—— and Maule, J. (eds), (1993) *Time Pressure and Stress in Human Judgment and Decision Making*. New York: Plenum.

Thill, J.-C. (1992) Choice set formation for destination choice modeling. *Progress in Human Geography*, 16, 361–82.

—— and Horowitz, J.L. (1997) Travel-time constraints on destination-choice sets. *Geographical Analysis*, 29, 108–23.

—— and Thomas, I. (1987) Toward conceptualizing trip-chaining behavior: A review. *Geographical Analysis*, 19, 1–17.

Thompson, D. (1963) New concept: Subjective distance – store impressions affect estimates of travel time. *Journal of Retailing*, 39, 1–6.

—— (1966) Future directions in retail area research. *Economic Geography*, 42, 1–19.

Thorndyke, P.W. and Hayes-Roth, B. (1982) Differences in spatial knowledge acquired from maps and navigation. *Cognitive Psychology*, 14, 560–89.

Timmermans, H.J.P. (1983) Non-compensatory decision rules and consumer spatial behavior: A test of predictive ability. *Professional Geographer*, 35, 449–55.

—— (1988) Multipurpose trips and individual choice behaviour: An analysis using experimental design data. In Golledge, R.G. and Timmermans, H.J.P. (eds), *Behavioral Modelling in Geography and Planning*. London: Croom Helm, pp.141–72.

—— and Golledge, R.G. (1990) Applications of behavioral research on spatial problems II: Preference and choice. *Progress in Human Geography*, 14, 311–54.

—— and van der Werden, P. (1992) Modelling sequential choice processes: The case of two-stop trip chaining. *Environment and Planning A*, 24, 1483–90.

Tversky, A. (1969) Intransitivity of preferences. *Psychological Review*, 76, 31–48.

—— (1972) Elimination by aspects: A theory of choice. *Psychological Review*, 79, 281–99.

Tversky, B. (1991) Spatial mental models. In Bower, G.H. (ed.), *The Psychology of Learning and Motivation: Advances in Research and Theory*, vol. 27, New York: Academic, pp. 109–45.

Verplanken, B., Aarts, H. and Van Knippenberg, A. (1997) Habit, information acquisition, and the process of making travel mode choices. *European Journal of Social Psychology*, 27, 539–60.

von Neumann, J. and Morgernstern, O. (1944) *Theories of Game and Economic Behavior.* New York: Wiley.

Warren, D.H. (1994) Self-localization on plane and oblique maps. *Environment and Behavior*, 26, 71–98.

5 Route learning and wayfinding

Edward H. Cornell and C. Donald Heth

Introduction

The nature of route learning seems straightforward. Parenting in modern societies provides an example. When a child is thought to be old enough to walk to school alone, a parent usually serves as a guide. The first path is chosen after leaving the home, but because the path may not be a direct line to the school, the parent points out landmarks that allow the child to know where to turn first. After that turn, the child follows another path, and the parent may point out a landmark to allow the child to monitor distance travelled. So it goes – the route to the school is simply a sequence of turns and paths. Because a route can be so easily demonstrated and described, and the use of routes is so common to our everyday commerce, explanations for the processes of route learning seem to be easy.

We now know that simple explanations are inadequate. In fact, the study of processes of route learning has led to the classic conundrums in psychology. What is the cue for turning – a landmark or a place? – has led to the question of how to define a stimulus. Whether actions are repeated or travel is directed to a goal has led to the question of what is the response. Whether a sequence of landmarks and turns is linked by associations has led to the question of what is learned. Why we are at a place has led to the question of the motivation of behaviour. How travel is distributed in the environment has led to the question of how space is represented.

In addition to expansion into these issues, it is now recognized that processes of route learning cannot be isolated from more general processes of wayfinding. Wayfinding is navigation that occurs both on and off known routes. Returning to our earlier example, it is completely natural that a child returning from school would step off the path walked with the parent. Distraction, curiosity, and adventure are characteristics of all travel, and there is growing evidence that 'lost' is a universal human experience (Hill, 1998; Montello, 1998a). If a new route is tried, or a portion of an old route is forgotten, the traveller is wayfinding.

Behavioural geographers suggest that route knowledge can be character-ized as a series of procedural descriptions (Golledge and Stimson, 1997).

The descriptions begin with an anchor point, a place that defines the beginning of the route. Route performance thereafter depends upon successful application of procedures that integrate appropriate sources of geographical information.

In wilderness parks, anchor points – such as trail heads – are places usually marked by notable landscape, ease of access, or signage. It would not be unusual that the markers for a wilderness route specify significant features of its course, as in 'Saddleback Pass', or anticipate the user's destination, as in 'Devil's Playground'. The procedural description for such a route may be simply to stay on a trail. The perceptual and cognitive processes for route learning would entail discrimination of the trail from bordering landscape, estimation of the duration and distance of travel, and identification of the end point. The route could be successfully retraced with no need for memories of the sequential order of events, although some serial memories occur incidentally and could be used to form expectations of scenes, bearings, and turns (Cornell *et al.*, 1999).

In contrast, in urban and suburban environments, paths and streets are extensively networked and there are usually several choices at the beginning and at intersections along a route. Different destinations are possible once a route has been initiated and the same destination can be reached by different routes. Hence, the procedural description for a route in the city usually necessitates a sequential record of events; in addition to the perceptual and cognitive processes listed above, re-using a route now requires choice point recognition, directional selection, path segment identification and ordering, and travel mode selection (Golledge and Stimson, 1997). In the city, it is especially easy to appreciate the chief advantage of using known routes over wayfinding. Known routes minimize cognitive effort; processes of wayfinding become simplified and automatic as a result of the repeated sequence of environmental events and associated actions.

This is not to say that different processes or different representations of spatial events are involved in route learning and wayfinding. Because people are constantly surrounded by spatial configurations, there is a growing appreciation that the elements of procedural knowledge are more than landmarks and turns. For example, it is clear that two processes usually associated with survey knowledge, bearing and distance estimation, are frequently used when making decisions along a route. A traveller may not recognize a building when looking down one of the streets at an intersection, but will select that street because it is approximately aligned with a bearing to the destination. The bearing may be characterized in a cardinal frame of reference, such as to the north, which can be inferred from the position of the sun at that time of day, memories of the direction of travel and turns from the start of the trip, or knowledge that streets in this segment of the route are generally aligned with the cardinal directions. In other words, although it is possible to follow a route by reacting to familiar landmarks as they are encountered, processes of wayfinding, world knowledge and survey knowledge seem to be inextricable components of route learning and are available during route reconstruction. The flexible and adaptive nature of spatial cognition en route is apparent across the research we review in this chapter.

The basic elements of routes

Comparative psychologists at the turn of the century had been challenged by popular accounts of foresight in what were supposedly forms of mental life that were lower than that of humans. Of particular interest was the notion that many creatures possessed an extraordinary sense of direction that allowed them to migrate vast distances. Naturalists suggested that some creatures appeared to minimize distance travel when they foraged for food, and some birds were reported to return accurately to caches that they had scattered about in obscure patterns. Were these travels instinctual or owing to route learning?

In one of the first studies directed to spatial problem solving by animals, Small (1901) examined the ability of rats to solve a miniature version of the Hampton Court maze. Maze studies rapidly gained in popularity, such that by 1935, Maier and Schneirla described the maze as 'the most important instrument that has yet been devised for animal studies' because, among other reasons, 'all higher animals learn their way about in nature'.

Much of this early research analysed performance in terms of route learning, in particular, how turns might be determined at choice points in the maze. Stimuli along the route, such as an odour or other landmark near an intersection of paths, were considered to be possible cues for actions, along with vestibular and kinaesthetic sensations. For example, rats were described as showing 'centrifugal swing' (Ballachey and Krechevsky, 1932) in which a forward momentum would predispose them to enter some alleys over others. This description suggested that, in addition to environmental cues, internal sensations of movement and effort were basic elements of route learning.

Another early discovery was that the choices made at the end of the maze were the earliest learned, with choices in the middle of the route learned the slowest (Maier and Schneirla, 1935). Interestingly, there are now several demonstrations that children and adults show the same pattern when learning routes in urban settings (Cornell *et al.*, 1989; Cornell *et al.*, 1992; Golledge *et al.*, 1985). An application has recently been demonstrated: The tendency for human wayfinders to err at intersections midway along a route can be used in police search operations to predict areas where lost persons may be found (Cornell *et al.*, 1996).

Hull (1932; Hull *et al.*, 1940) constructed a unified account of reinforcement effects on positional responses that explained the serial order by which turns and continuations along a route were learned. The end point or goal of a maze included the reinforcing event for a chain of responses.

Earlier choices were further away from the reinforcer in time and therefore were learned more slowly. However, early choices also were less susceptible to interference from memories from previous choices. The result was a predicted pattern of route learning, in which choices in the middle of the route would be learned more slowly.

Hull *et al.*'s (1940) characterization of how responses might be linked to cause the good performance near the end points of routes appears to be an ancestor of modern connectionistic networks. Connectionistic networks are models of the connections between processing units analogous to neurons; the presentation of a pattern of inputs leads to a pattern of activation of output nodes that represents a category of knowledge or action. Note, however, that Hull's attempt to show the mutual influences of associations learned along a route did not include higher level nodes. In connectionist networks, higher level nodes integrate or represent the interactions among the responses of lower level processors. Hence, Hull's model did not include a mechanism that could easily accommodate the fact that travellers differentiate segments of routes.

In sum, early theories of maze learning emphasized a linear sequence of correct choices that brought the animal closer to the location of the maze incentive. However, it was soon discovered that animals could learn various stimuli when making choices along a route, and the adaptive nature of their spatial problem solving suggested generalized systems of behaviour. For example, using the field model of social psychology described by Lewin (1938), Tolman (1932) suggested a role for holistic spatial stimuli in governing simple behaviours in mazes. Rats would, for example, resist pathways that took them in redundant loops (Shepard, 1931) and would, conversely, seek shortcuts or detours when faced with blockades of normal routes (Tolman and Honzig, 1930). Tolman's theoretical initiative to explain these behaviours was the well-known 'cognitive map', which he generalized to a wide variety of human psychological phenomena (Tolman, 1948).

The popularity of this metaphor, however, should not obscure the success that other behaviourists achieved by using more response-based accounts. Most of the alternatives to Tolman's cognitive map built on the elements of route learning suggested by Hull and his associates. For example, Spence (1956) used the concept of anticipatory goal responses to construct an elaborate chain of covert behaviours that would represent an organism's experience with a given response sequence and would, in a behaviouristic sense, provide an expectation of overt turns. Similarly, Deutsch and Clarkson (1959) extended the behaviourist model to explain some forms of wayfinding and detour-taking. Configurations of landmarks and distant landmarks could serve to indicate bearings when specific local cues were unavailable.

In retrospect, it seems that the attempts by early experimental psychologists to explain maze performance fostered a preternatural separation of elements. Hull and his followers demonstrated that chains of stimulus-response connections were elements of route learning, but it is now known that these elements cannot easily represent the areal and configurational qualities of experience along a route (Hilgard and Bower, 1966). Tolman and his followers demonstrated that bearing and distance were elements of route learning, but it is now known that these elements are not represented mentally as they would be in a survey map (cf., Hirtle and Heidorn, 1993; Kuipers, 1983). As we shall see, both forms of elements are necessary in contemporary descriptions and applications.

Place recognition along routes

Early analyses suggested that landmarks were stimuli external to the animal that could serve as cues for action. Examples of landmarks were a marking on the path, an odour down an alley, or a house light visible over the walls of the maze. It was soon realized that this conception of landmarks was overly reductionistic. To be a landmark, an object must be part of a spatial relationship; a landmark is always defined in relation to another object or frame of reference (Blades, 1989; Presson and Montello, 1988). Because our sense systems register the spatial configuration of events, the place that situates a landmark serves as an immediate frame of reference. For example, when scenes are viewed by humans, objects can be sensed in peripheral vision as a central landmark is fixated. With a short eye movement, these objects can be identified and seen to be neighbours. Nevertheless, because of limits on our perceptual and memory systems, the spatial relations of multiple objects may not be immediately apprehensible.

Young children especially have difficulty learning the configuration of a place. An early study of recognition of pictures of natural scenes showed that the difficulty is not that young children focus their attention on a salient object exclusively (Kirasic *et al.*, 1980). When landmarks in the centre of scenes remained the same, five-year-olds detected changes that had been made in surrounding objects. When the objects in the surround remained the same, five-year-olds detected changes that had been made in central landmarks. However, in response to both kinds of foils, five-year-olds did not recognize scenes as accurately as nine-year-old children and twenty-two-year-old adults. Analyses of the time taken to make recognition judgements suggested that the young children's difficulty was that they were slow in encoding the relations between landmarks and their context.

This result suggests that one of the chief developments in the ability to recognize places is speed of processing. As fixated objects are more quickly identified and localized, more flanker objects can be fixated (Rayner and Pollatsek, 1992). The ability to register what is peripheral allows a place to be known with more area. There are implications for route learning as well. Younger children typically know less than older children and adults about the sequence of events along a newly acquired route (Siegel *et al.*, 1978). Gaps in route knowledge could occur, for example, if a younger child learned a landmark at an intersection, but the spatial extent of his or her attention to nearby landmarks was less than that of older children.

This interpretation is supported by an analysis of the landmarks that children named when they actually returned along a route (Heth *et al.*, 1997). Each of four intersections was photographed to include 35° of visual field centred on a landmark that had been pointed out to the child during the initial walk along the route. A landmark that had been named at these intersections was considered close if it could be identified within the picture and distant if it could not. Of the mean of 6.1 landmarks named by eight-year-old children, 26 per cent were distant; of the mean of 8.0 landmarks named by twelve-year-old children, 38 per cent were distant.

Interestingly, there were also qualitative differences in the landmarks named by the different age groups. Adults judged whether the landmarks named by the children were stationary, or likely to be in the same place when returning along the route. Of the landmarks named by eight-year-old children, 51 per cent were stationary; of the landmarks named by the twelve-year-old children, 62 per cent were stationary. Note that more than speed of processing is required to select a landmark that adults judge to be stationary. Children need to acquire world knowledge – or experience that reliable landmarks do not have wheels, are not animated, are massive, or rooted – to designate places along a route.

When particular landmarks cannot be remembered, routes can also be known by the relative familiarity of places as they are encountered (Cornell *et al.*, 1994). For example, consider how a wayfinder can repeat a route from beginning to end point. The first intersection would involve a choice of paths, and the uncertain wayfinder could pause to compare the alternatives. For one or more of the alternatives, landmarks along the path should be unfamiliar because previous travel did not provide close or extended experience with those landmarks. Landmarks along one of the alternative paths should be familiar because they had been seen both in the background and in the foreground as their place was approached. In addition, because the visual field of the traveller is usually oriented toward the direction of movement, landmarks in the front received more exposure than landmarks in the periphery or rear during the original experience along the route. These observations suggest that the strongest feeling of knowing would occur when the original path is viewed in the original direction of travel. Hence, whenever a decision is required for directing travel from the beginning to the end point of a route, the correct choice is to go toward the route features that yield the strongest memories.

The analysis can be extended to the problem of returning along a newly learned route as well (Cornell *et al.*, 1994). The advantage of wayfinding by approaching familiar places is that directional choices do not require a map-like representation, or a list of landmarks and associated actions, or even expectancies of landmarks. The method could be especially useful when a traveller has not completely acquired or structured spatial or serial information about a route. Young children who lack cognitive abilities required for organized spatial representations may rely on judgements of the familiarity of places for navigation (Blades and Spencer, 1987).

However, a slow rate of encoding of neighbours of landmarks would be likely to reduce place recognition and navigation accuracy. For example, after being led astray while returning along a route, eight-year-old children often failed to judge they were off route when they could see intersections that they originally viewed from on route (Cornell, *et al.*, 1994). More often than twelve-year-old children or twenty-five-year-old adults, the eight-year-old children erroneously proceeded at these intersections when they could have selected the path that allowed them to return on route.

Hence, more strategic kinds of navigation may be advantageous when familiarity judgements are difficult. Before discussing what different navigation strategies reveal about the nature of routes, we describe how route knowledge progresses from episodic memories of places.

The segmentation of routes

Routes in flat, homogenous environments can bear directly toward a target. In circumstances such as expansive desert, open sea, or prairies of fresh snow, there are no paths worn on the earth's surface, and a star on the horizon may serve as a beacon to approach, or the direction of prevailing wind, magnetic force, or ocean current may serve as a line to estimate a course. While maintaining a bearing across an undifferentiated surface, the sensations of time and motion may be the only means to estimate progress.

In contrast, routes in heterogeneous environments traverse through patches of landscape, circumventing barriers and obstacles, usually following the corridors between them, but sometimes climbing over or dropping beneath. Routes in heterogeneous environments are rich with information. Changes in surroundings help to define a sequence. Places along the route – such as the location of a spring – may be distinguished by objects, scenes, or sounds that do not occur at other places along the route. And a pattern of phenomena such as a confluence of paths, a thickening of greenery, or the revelation of a panorama can mark such an area (Heft, 1983). Golledge (1976) first suggested that these environmentally distinguished places are the basis of nodes in the representation of routes.

Why would nodes be useful? Earlier, Carr and Schlisser (1969) had suggested that without some means of organizing perceptual intake along a heterogeneous route, route learners would soon be overloaded and confused. Carr and Schlisser further suggested that the ability to recognize the structural properties of sequences of events, their regularities and recurrence, provided a natural means to segment routes. Segmentation, like other forms of grouping of related information, would reduce the number of memories to be apprehended when the route is recalled. There are fewer segments than memories of events along the route, so the serial order of segments should be easier to learn. Nodes could provide memorable end points for segments of routes, even segments that contained landmarks that were poorly encoded or forgotten.

This analysis of human conceptions of routes was neatly sustained in experiments by Allen (1981). Elementary school children and college students viewed sixty photographs ordered to depict a walk that began in a wooded park, continued across a university campus, and ended in a residential area. When asked to show how many different parts there were to the walk, students partitioned the sequence of photographs into route segments. The boundaries for these segments were the same for all age groups and consisted of photographs taken at transitions between environmentally distinguished portions of the route. For example, scenes depicting the campus were placed in a row and scenes depicting residential housing constituted another row; scenes depicting a major street bounding these two areas were selected as the last scene of the campus row and the first scene of the residential row. Further, when different students of the same ages were asked to estimate the walking distance between scenes along the route, their proximity judgements showed systematic errors. Proximity judgements for pairs of scenes selected within segments were more accurate than proximity judgements for pairs of scenes selected across different segments (cf., Cohen *et al.*, 1978).

These results indicate that routes are represented by more than lines. Routes are also hierarchical, with landmarks, bearings, and actions embedded within segments, which are embedded within a larger spatio-temporal framework, defined by the beginning and end of the trip. Allen (1981) suggested that relations between landmarks in segmented routes were topological; that is, the representation of distance between landmarks is not based on a metric, but is biased by whether the landmarks are contained or enclosed within the same segment. However, Montello (1998b) has recently presented compelling arguments that both metric and topological representations are involved in estimating the extent of routes.

The layout of routes

In a grand analysis, a person's movements throughout life could be represented as nodes, segments, and routes (Golledge, 1978). Such a record would also have to indicate spatial relations between routes, which could be mapped as a skeletal structure, or network. This network would include configurational relations that are characteristic of a survey representation, but the concatenated routes would not have to be situated relative to a superordinate frame of reference. For example, it is not necessary to characterize route bearings in relation to cardinal directions. A route can be said to consist of a narrow path bearing off of a major thoroughfare, the second path could be described as oriented toward a feature of the skyline, or the route can be characterized egocentrically by a series of actions such as a gradual right turn followed by a hard left.

Evidence suggests, however, that as a result of both perceptual and cognitive processes, linear and skeletal representations of routes are embedded in areal representations (Gärling *et al.*, 1982; Evans *et al.*, 1981; Lindberg and Gärling, 1982). Typical travel reveals geometrically determined perspectives of distant scenes and tall landmarks from different sites along a route. By peering down intersecting paths or pausing to explore at open viewpoints, travellers may directly perceive Euclidean relations between routes (Gibson, 1979). It seems likely that processes of direct perception could be the basis for inferences about how routes are situated within macro-spatial areas or formal frames of reference (Rieser *et al.*, 1986).

For example, Moar and Carleton (1982) suggest that, if a route in a city reveals a grid-block layout, wayfinders may assume that nearby routes are straight and are positioned at right angles to one another. In many cities, major routes and grids are aligned with cardinal directions. Moreover, there is evidence that route learners can integrate their experiences within a configural representation even when distant landmarks or regularities of layout are not visible from separate routes (Montello and Pick, 1993).

The situation providing this evidence is intriguing. College students learned landmarks along two separate routes when they were led through a large building complex containing several levels. The two routes were located one above the other, but participants remained unaware of this because they walked through a tunnel between the starting points of the routes. After learning the second route, participants were led to one of the starting points and informed that the place they were standing was in fact above or below the starting point for the other route. Participants were permitted some time to think about this revelation, then asked to imagine where the other route and all the landmarks were located from where they currently stood. The latency and accuracy to point to imagined landmarks was impaired when pointing from one route to the other as compared to pointing within a route. Nevertheless, the magnitude of directional error both between and within routes was significantly less than chance, indicating good acquisition and integration of configurational knowledge.

These results support an earlier assertion by Moar and Carleton (1982): when route learners see from separate routes the same objects in the skyline, or otherwise become aware of a shared frame of reference, they are immediately able to construct a good estimate of the spatial layout of the routes.

Route ecology

Earlier we saw that people can approach familiar places to find their way along routes. The analysis suggested that perceived configurations of objects define places, which can be remembered as nodes in larger serial and spatial representations. Here, we consider how more strategic navigation processes inform us about the representation and use of routes. We focus on strategies that have probably characterized human solutions from the time we began hunting and gathering.

The rapid expansion of *homo erectus* in the late Pleistocene era brought humans into vastly different environments with special challenges to wayfinding. Although artifacts from the scattered campsites of these early humans have occasionally permitted hypotheses about cognitive functioning (Donald, 1991), there are few hints regarding how routes were learned. Hunter-gatherer societies still exist, however, and anthropological descriptions suggest that route learning is substantially affected by the ecology and geomorphology of the environment.

A helpful distinction in this regard is that used by Gallistel (1990), who describes differences between *piloting* and *dead reckoning*. Piloting is movement controlled by the constant monitoring of stable environmental features. A distinctive peak may be kept in view as a trail courses through mountainous corridors. More formally, piloting uses azimuthal fixes whose intersection provides a location. In contrast, dead reckoning (a corruption of the expression 'deduced reckoning') monitors velocity and direction cues to infer a position. As practised by early navigators, for example, the relative motion of a ship against the current was used to infer velocity, which combined with time, gave distance estimates. Angular information was gathered from wind and wave tracks (later superseded by compass bearings) to provide direction. Although piloting and dead reckoning are combined in formal navigation systems, Gallistel notes how the information available in the organism's niche will determine which is a feasible alternative for wayfinding.

The Inuit people of the arctic are adept at using dead reckoning during hunting excursions. Although variations in ice features can provide some natural landmarks, they are frequently unreliable. Winds and currents move the sea ice relative to the land, and the weather of the region may occasionally obscure visibility beyond a few metres. Furthermore, specific trails are often rendered useless, either by the movement and rotation of the ice pack, or by the opening of a large, impassable channel of open water across the path of travel.

Nelson (1969) describes how Inuit hunting expeditions are frequently faced with the need for detours and altered plans. Expert hunters seem to use those environmental features which provide clues to direction and distance without providing absolute location. The angle of the wind on a hunter's face or the intersecting angle of the trail with the snow drifts, for example, provide clues to direction, while estimates of travel speed and time allow the hunter to judge distance. In addition, hunters employ heuristics that naturally correct possible errors of dead reckoning, such as the use of 'catch features', in which a hunter deliberately overestimates a heading to ensure that he is not short of a planned goal. The use of a dead reckoning system seems well suited to an environment with fluid geography and large expanses of changing features.

In contrast, aboriginals of the vast Western Australian desert live in a more static environment containing important punctate resources such as waterholes and patches of special vegetation. Despite their stability, the locations of these resources are obscure to outsiders. Flat earth, scrub trees and clusters of rocks appear to be indistinguishable. Provided that landmarks can be differentiated, this environment permits navigation strategies more akin to piloting. In fact, aboriginal travel narratives are described as an ability to 'read that country' (Sansom, 1980). Memories of permanent information are intrinsic to walking about.

Routes and place names are remembered and communicated through ritual song cycles characterized as 'oral maps' (Davis and Prescott, 1992). These songs recount mythic voyages by the totemic beings identified with a given group or clan. The activities of these beings are part of ancestral creation myths and are identified with specific geographic locations. The place where the-snake-fell-asleep-on-the-warm-rock is a location that could be concatenated with the place where the-snake-changed-into-a-cat-and-drank-from-the-river. As a result, journeys of ancestral beings form an extensive series of elaborated events. Hierarchical structure is also common, as individual sites are combined into 'big place' names (Berndt, 1976).

The cultures of the Western Australian desert show highly particularized instances of these spatial representations. Individual clans maintain ownership of song sequences, which, in turn, convey ownership of the locations named in the journey. The sequence describes a 'track' which recapitulates the origin myth of the group. Tindale (1972) reproduces one such track, sketched by a member of one of the Pitjandjara groups. The depiction is strongly reminiscent of Tolman's (1948) strip maps, with long linear segments and highly regularized turns. Although apparently organized as a network, sometimes these tracks will be discontinuous, with gaps between major segments (Davis and Prescott, 1992). The area associated with a gap may not be known because it belongs to another clan, and the gaps are described as places where their own ancestral being travelled underground, or through the air, to reappear in another locale. Segment boundaries are often marked by changes in the language of the place name, which is interpreted to mean ownership by another group.

Seasonal movement by a clan or group of the Western Australian desert culture is strongly influenced by these tracks. Although watering places might permit a direct line of travel based upon dead reckoning, a group's route will, instead, take place along the track described by the group's ancestral myths (Berndt, 1972). Young males learn the idiosyncratic track during ritual initiation ceremonies in which they are taught the myths and shown the locations of historic events. Among the Pitjandjara, for example, a boy who has been recently engaged in marriage will participate in dance and song ceremonies culminating in his circumcision. He is then taken, under a ban of silence, on a tour of ceremonial places and instructed as to their totemic significance. This period may last as long as a year, and include three hundred sites (Tindale, 1972). The period culminates with additional ceremonies and further genital scarring. The narrative formats, elaboration through re-enactment, ritualistic repetition, and physiological arousal provide extensive mnemonics for unchanging sequences of route events.

The modern city contains ecological constraints similar to both of the environments we have described. Like the Western Australian desert, an urban landscape contains many consistent cues as well as natural channels and habitual tracks. However, like the open ice channels and blizzards of the Arctic ecology, dynamic factors such as traffic and construction can impede planned routes in the city.

Knowledge of routes in the modern city seems to reflect these functional requirements. In an extensive study of taxi drivers in Pittsburgh, Pennsylvania, Chase (1983) discovered that expert urban navigation did not proceed by reference to a 'map in the head'. Expert drivers could quickly name the shortest routes between two locations and could also quickly name the shortest detour around a barrier. Knowledge of the street system seemed to be hierarchically organized, so that major thoroughfares were used to interconnect neighbourhoods, and, within neighbourhoods, secondary streets were used to connect the origin and destination of the trip with the thoroughfares. Neighbourhoods were known to exist within regions of the city, and large regions were known with respect to more global features, such as a river junction downtown.

However, the hierarchical representations and reliance on routes by the expert taxi drivers were associated with an unexpected finding: Driving throughout the city for ten or more years did not produce accurate survey knowledge. Neighbourhoods were thought to be closer to the downtown area than they were, and when asked to point to downtown locations, the drivers' pointing was biased by the orientation of the nearest street that lead downtown. Distances were overestimated when they involved locations in different neighbourhoods, and overestimations were even larger when the neighbourhoods were separated by a physical barrier such as a river. In addition, taxi drivers were able to improve on their descriptions of routes when they were in the field. They generated better routes as they encountered landmarks and boundaries, indicating that their network knowledge was perceptually cued (Chase, 1983).

These ecological descriptions belie a stage model where routes are comprised of topological knowledge intermediate between memories of landmarks and survey representations (Hart and Moore, 1973; Shemyakin, 1962; Siegel and White, 1975). Routes for Arctic hunters can be survey-based vectors, calculated during travel as angular displacements from sightings of celestial bodies (MacDonald, 1998). Routes for Australian bushmen can be narratives and songs, with gaps, elaborations, and hierarchies atypical of seriated lists. Routes for Pittsburgh taxi drivers can be segments of a network, but vector distance is not accurate and the layout of regions is subordinate to the traversible segments that connect them (cf. Byrne, 1979; Chase, 1983). Hence, the geography and function of travel determine how routes are learned.

Further on up the road

Several puzzles remain concerning the processing of route information. Attempts by children and adults to find alternatives to a familiar route are particularly fascinating. What has to be known about a route to take a shortcut? What has to be known about a route to take a detour?

The simplest shortcut may be one that is precipitated by spotting a landmark that appears beyond an upcoming turn. Approaching the landmark by line-of-sight usually takes the wayfinder across new territory efficiently and safely. Hence, a line-of-sight shortcut may only require recognition of landmarks alongside a familiar route. In contrast, a shortcut across territory that does not provide views of a familiar route may require memories of a landmark in the skyline. A direct approach toward the distant landmark may not lead to the intended destination, but perspective views of the landmark may be used to direct the shortcut to one side or the other. Finally, there is the process that has intrigued researchers since Tolman's original demonstrations, in which the inference of a shortcut requires a mental representation of the configuration of the route as if viewed from a survey perspective. Construction of this representation may begin by a wayfinder keeping track of the direction and extent of travel along a route segment with respect to the point of origin. Then, following a turn, the wayfinder may keep track of the direction and extent of travel with respect to the point of origin of the route, other sites along the first route segment, or landmarks in the distance that were seen during travel from the perspectives of the first route segment. One powerful hypothesis is that information about the layout of these events can be directly perceived – for example, the pattern of optical flow of objects and textures is correlated with the direction and extent of movement (Gibson, 1979). As they travel, sighted persons may use their experience with optical flow to keep track of even out-of-sight locations (Rieser et al., 1995). In addition, while walking up hills, sighted persons can see transitions from frontal views of objects at the horizon to top views of those same objects at an altitude. Hence, the core characteristic of cognitive map – survey representation – may be derived from universal experiences of survey viewing. After imagining a bird's-eye view of the layout of events experienced along a route, wayfinders could connect places along the route with a least distance line.

Each of these process descriptions of shortcuts needs to be elaborated by studying humans navigating real world routes. A complementary problem is how travel off route is directed when detours are required. Detours are usually longer and more indirect than the familiar route to a destination.

Detours may even require navigation beyond the margins of the cognitive map. A good understanding of off route navigation may inform police search operations for persons who become lost (Cornell *et al.*, 1996).

One of the most profound implications of the study of route learning was suggested by Arthur and Passini (1992). They pointed out that environmental design needs to consider ease of wayfinding. Visitors become frustrated trying to find offices in complex government buildings, take up the time of medical professionals when they ask directions in hospitals, and choose not to shop at certain malls when they want to go directly from their car to pick up an item. Early analyses of indoor wayfinding suggested that signage and colour codes could provide landmarks, but the addition of these cues after construction can be futile (Passini and Shiels, 1987). Designers who have considered the accessibility of destinations and the configuration of paths have arranged more legible routes. However, we are optimistic that, as the processes of exploration during route learning are understood, environments and trails will be more often planned to allow adventure and serendipity as well as efficiency. Windows affording access to external frames of reference – views of distant landmarks or natural boundaries will more often be available near choice points indoors. Short paths to alcoves and viewpoints – places for privacy and reflection will more clearly branch, so that the wayfinder on a mission on the main trail will not be distracted.

We anticipate that the study of route learning will also lead to improvements in maps. Case histories of persons lost in wilderness areas have indicated that route maps in guide books, you-are-here postings, or pamphlets available at visitor centres were sometimes inadequate (Heth and Cornell, 1998). These maps gave distances and showed the configuration of trails, but had little information about the area that contained the trails. Topography was not represented, so mountain bikers often continued on a down hill run rather than turning up a path at an intersection. Some distant landmarks were not represented, so that hikers who could not translate angles of intersecting lines from map to trail could not use a prominent mountain to infer their bearing. Our review has suggested that, from the onset of travel, route learning involves a variety of available spatial information. After drawing the line, the cartographer could represent for the map reader what needs telling from the perspective of the route.

Finally, it was not so long ago that futurists believed that we could create autonomous wayfinding robots. The attempts to do so have stalled, and the study of route learning indicates that, even if problems with landmark recognition are solved, higher order knowledge such as the configuration of routes would be necessary to produce correct responses outside of built environments. It would be interesting to observe a robot that makes decisions based on both a list of landmark-turn sequences along a path and consistency of action with a bearing toward a distant landmark (Chown *et al.*, 1995). Such a robot may be able to make detours or reach a destination while off a programmed route. However, the practical trend in the design of intelligent solutions is to allocate machine resources where humans fail, as in the retrieval of memories and rapid calculations, and allocate human resources where machines fail, as in complex pattern recognition, rapid learning, and adaptation of strategies. Hence, we predict that robots used for extra-terrestrial routes will continue to have a camera linked to a human navigator.

Acknowledgements

Our research is supported by grants from the Natural Sciences and Engineering Research Council and the Search and Rescue Secretariat of Canada.

References

Allen, G.L. (1981) A developmental perspective on the effects of 'subdividing' macrospatial experience. *Journal of Experimental Psychology: Human Learning and Memory*, 7, 120–32.

Arthur, P. and Passini, R. (1992) *Wayfinding: People, Signs, and Architecture.* Toronto: McGraw-Hill Ryerson.

Ballachey, E.L. and Krechevsky, I. (1932) 'Specific' versus 'general' orientation factors in maze running. *University of California Publications in Psychology*, 6, 83–97.

Berndt, R.M. (1972) The Walmadjeri and Gugadja. In Bicchieri, M.G. (ed.), *Hunters and Gatherers Today.* Toronto: Holt Rinehart Winston.

—— (1976) Territoritality and the problem of demarcating sociocultural space. In Peterson, N. (ed.), *Tribes and Boundaries in Australia.* Atlantic, NJ: Humanities.

Blades, M. (1989) Children's ability to learn about the environment from direct experience and from spatial representations. *Children's Environments Quarterly*, 6, 4–14.

—— and Spencer, C. (1987) Young children's wayfinding: The pedestrian use of landmarks in urban environments and on maps. *Man-Environment Systems*, 17, 105–12.

Byrne, R.W. (1979) Memory for urban geography. *Quarterly Journal of Experimental Psychology*, 31, 147–54.

Carr, S. and Schlisser, D. (1969) The city as a trip. *Environment and Behavior*, 1, 7–35.

Chase, W.G. (1983) Spatial representations of taxi drivers. In Rogers, D.R. and Sloboda, J.A. (eds), *Acquisition of Symbolic Skills.* New York: Plenum.

Chown, E., Kaplan, S. and Kortenkamp, D. (1995) Prototypes, location, and associative networks (PLAN): Towards a unified theory of cognitive mapping. *Cognitive Science*, 19, 1–51.

Cohen, R., Baldwin, L.M. and Sherman, R.C. (1978) Cognitive maps of a naturalistic setting. *Child Development*, 49, 1216–18.

Cornell, E.H., Heth, C.D. and Alberts, D.M. (1994) Place recognition and way finding by children and adults. *Memory and Cognition*, 22, 633–43.

Cornell, E.H., Heth, C.D. and Broda, L.S. (1989) Children's way finding: Response to instructions to use environmental landmarks. *Developmental Psychology*, 25, 755–64.

——— , ——— , Kneubuhler, Y. and Sehgal, S. (1996) Serial position effects in children's route reversal errors: Implications for police search operations. *Applied Cognitive Psychology*, 10, 301–26.

——— , ——— and Rowat, W.L. (1992) Way finding by children and adults: Response to instructions to use look-back and retrace strategies. *Developmental Psychology*, 28, 328–36.

——— , ——— and Skoczylas, M.J. (1999) The nature and use of route expectancies following incidental learning. *Journal of Environmental Psychology*, 209–29.

Davis, S.L. and Prescott, J.R.V. (1992) *Aboriginal Frontiers and Boundaries in Australia*. Carleton, Victoria: Melbourne University Press.

Deutsch, J.A. and Clarkson, J.K. (1959) Reasoning in the hooded rat. *Quarterly Journal of Experimental Psychology*, 11, 150–4.

Donald, M. (1991) *Origins of the Modern Mind: Three Stages in the Evolution of Culture and Cognition*. Cambridge, MA: Harvard.

Evans, G.W., Marrero, D.G. and Butler, P.A. (1981) Environmental learning and cognitive mapping. *Environment and Behavior*, 13, 83–104.

Gärling, T., Böök, A. and Ergezen, N. (1982) Memory for the spatial layout of the everyday physical environment: Differential rates of acquisition of different types of information. *Scandinavian Journal of Psychology*, 23, 23–35.

Gallistel, C.R. (1990) *The Organization of Learning*. Cambridge, MA: MIT.

Gibson, J.J. (1979) *The Ecological Approach to Visual Perception*. Boston, MA: Houghton Mifflin.

Golledge, R.G. (1976) *Cognitive Configurations of a City*, vol. 1. Columbus, OH: Ohio State University Research Foundation.

——— (1978) Learning about urban environments. In Carlstein, T., Parkes, D. and Thrift, N. (eds), *Timing Space and Spacing Time: I. Making Sense of Time*. London: Edward Arnold, pp. 76–98.

——— Smith, T.R., Pellegrino, J.W., Doherty, S. and Marshal, S.P. (1985) A conceptual model and empirical analysis of children's acquisition of spatial knowledge. *Journal of Environmental Psychology*, 5, 125–52.

——— and Stimson, R.J. (1997) *Spatial Behavior: A Geographic Perspective*. New York: Guilford.

Hart, R.A. and Moore, G.T. (1973) The development of spatial cognition: A review. In Downs, R.M. and Stea, D. (eds), *Image and Environment: Cognitive Mapping and Spatial Behavior*. Chicago, IL: Aldine, pp. 246–88.

Heft, H. (1983) Way-finding as the perception of information over time. *Population and Environment*, 6, 133–50.

Heth, C.D. and Cornell, E.H. (1998) Characteristics of travel by persons lost in Albertan wilderness areas. *Journal of Environmental Psychology*, 223–35.

——— , ——— and Alberts, D.M. (1997) Differential use of landmarks by 8- and 12-year-old children during route reversal navigation. *Journal of Environmental Psychology*, 17, 199–213.

Hilgard, E.R. and Bower, G.H. (1966) *Theories of Learning*, 3rd edn, New York: Appleton Century Crofts.

Hill, K. (1998) The psychology of lost. In Hill, K. (ed.), *Lost Person Behavior*. Denver, CO: National Association of Search and Rescue.

Hirtle, S.C. and Heidorn, P.B. (1993) The structure of cognitive maps: Representations and processes. In Gärling, T. and Golledge, T. (eds), *Behavior and Environment: Psychological and Geographical Approaches*. Amsterdam: Elsevier/North-Holland.

Hull, C.L. (1932) The goal-gradient hypothesis and maze learning. *Psychological Review*, 39, 25–43.

—— Hovland, C.I., Ross, R.T., Hall, M., Perkins, D.T. and Fitch, F.B. (1940) *Mathematico-Deductive Theory of Rote Learning.* New Haven, CT: Yale.

Kirasic, K.C., Siegel, A.W. and Allen, G.L. (1980) Developmental changes in recognition-in-context memory. *Child Development*, 51, 302–5.

Kuipers, B. (1983) The cognitive map: Could it have been any other way? In Pick, H. and Acredolo, L. (eds), *Spatial Orientation: Theory, Research, and Application*. New York: Plenum.

Lewin, K. (1938) The conceptual representation and the measurement of psychological forces. *Contributions to Psychological Theory*, 1. Durham, NC: Duke.

Lindberg, E. and Gärling, T. (1982) Acquisition of locational information about reference points during locomotion: The role of central information processing. *Scandinavian Journal of Psychology*, 23, 207–18.

MacDonald, J. (1998) *The Arctic Sky: Inuit Astronomy, Star Lore, and Legend.* Toronto: Royal Ontario Museum.

Maier, N.R.F. and Schneirla, T.C. (1935) *Principles of Animal Psychology.* New York: Dover.

Moar, I. and Carleton, L.R. (1982) Memory for routes. *Quarterly Journal of Experimental Psychology*, 34, 381–94.

Montello, D.R. (1998a) A new framework for understanding the acquisition of spatial knowledge in large-scale environments. In Golledge, R. and Egenhofer, M. (eds), *Spatial and Temporal Reasoning in Geographic Information Systems*. New York: Oxford University Press.

—— (1998b) What it means to be lost. Proceedings of the Search and Rescue Secretariat of Canada (SARSCENE), Banff, Alberta.

—— and Pick, H.L. (1993) Integrating knowledge of vertically aligned large-scale spaces. *Environment and Behavior*, 25, 457–84.

Nelson, R.K. (1969) *Hunters of the Northern Ice.* Chicago, IL: University of Chicago Press.

Passini, R. and Shiels, G. (1987) *Wayfinding in Public Buildings: A Design Guideline* (Document AR56b). Ottawa, Ontario: Public Works Canada.

Presson, C.C. and Montello, D.R. (1988) Points of reference in spatial cognition: Stalking the elusive landmark. *British Journal of Developmental Psychology*, 6, 378–81.

Rayner, K. and Pollatsek, A. (1992) Eye movements and scene perception. *Canadian Journal of Psychology*, 46, 342–76.

Rieser, J.J., Guth, D.A. and Hill, E.W. (1986) Sensitivity to perspective structure while walking without vision. *Perception*, 15, 173–88.

—— Pick, H.L., Ashmead, D.H. and Garing, A.E. (1995) Calibration of human locomotion and models of perceptual-motor organization. *Journal of Experimental Psychology: Human Perception and Performance*, 21, 480–97.

Sansom, B. (1980) *The Camp at Wallaby Cross: Aboriginal Fringe Dwellers in Darwin.* Atlantic, NJ: Humanities.

Shemyakin, F.N. (1962) General problems of orientation in space and space representation. In Anan'yev, B.G. (eds), *Psychological Science in the USSR, Vol. 1, Tech. Report No. 11466*. Washington, DC: US Office of Technical Services, pp. 184–255.

Shepard, J.F. (1931) More learning. *Psychological Bulletin*, 28, 240–1.

Siegel, A.W., Kirasic, K.C. and Kail, R.V. (1978) Stalking the elusive cognitive map: The development of children's representations of geographic space. In Altman, I. and Wohlwill, J.F. (eds), *Human Behavior and Environment*, vol. 3. New York: Plenum.

—— and White, S.H. (1975) The development of spatial representations of large-scale environments. In Reese, H. (ed.), *Advances in Child Development and Behavior*, vol. 10. New York: Academic, pp. 9–55.

Small, W.S. (1901) An experimental study of the mental processes of the rat. II. *American Journal of Psychology*, 12, 206–39.

Spence, K.W. (1956) *Behavior theory and conditioning*. New Haven, CT: Yale University Press.

Tindale, N.B. (1972) The Pitjandjara. In Bicchieri, M.G. (ed.), *Hunters and Gatherers Today*. Toronto: Holt Rinehart Winston.

Tolman, E.C. (1932) *Purposive Behavior in Animals and Men*. New York: Appleton-Century.

—— (1948) Cognitive maps in rats and men. *Psychological Review*, 55, 189–208.

—— and Honzig, C.H. (1930) 'Insight' in rats. *University of California Publications in Psychology*, 4, 215–32.

6   Understanding and learning maps

Robert Lloyd

For the truth is that the unravelling of many of the mysteries of cartographic design and presentation has not yet been accomplished.

(Robinson, 1952)

Introduction

Maps are two- or three-dimensional spatial structures that represent some
part of the environment and communicate information about the environment. Maps differ from photographs of the environment in that someone has
organized the spatial data and represented it as spatial information to be communicated. The cartographer (map maker) displays the information and the
map reader acquires the information (Figure 6.1). The spatial information on
the map becomes spatial knowledge when patterns in the information are
learned by the map reader. The hypothetical process illustrated in Figure 6.1
shows the initial data as numbers and the learned knowledge as connections
in a neural network. The numbers, of course, are measurements of the
environment, the real world, and beyond the neural connections is the meaning the spatial knowledge has for the cognitive mappers.

Cartographers refer to this process as the cartographic communication
model (Board, 1967; Koláčny, 1969). A map can be thought of as an expression of a cartographer's ideas, a device for storing spatial information and
a source of knowledge for the map reader. Some fundamental changes occur
when the cartographer transforms spatial data into spatial information and
other changes occur when the map reader transforms spatial information
into spatial knowledge (Figure 6.1). In a sense, cognitive maps that are
learned from cartographic maps are second derivatives of the environment.
Cartographers categorize, generalize, and symbolize to enhance important
information and eliminate any non-essential information. Cognitive mappers
continue this simplification process when they encode information from the
cartographer's map. Most cognitive maps not only fail to reflect all the
details of the environment they represent, but also have systematic errors
caused by the processes that encode them into memory (Tversky, 1981,
1992).

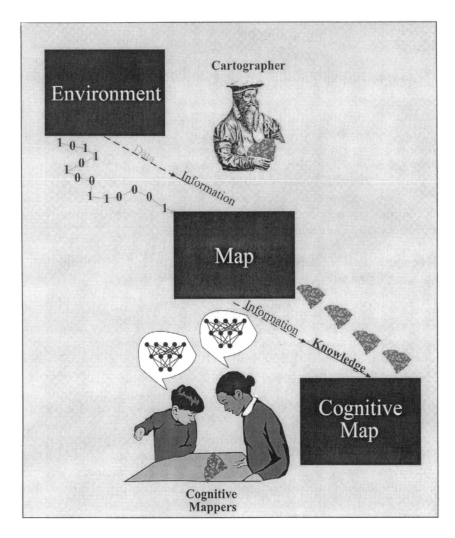

Figure 6.1 The cartographic communication process. The cartographer transforms spatial data into spatial information and cognitive mappers transform spatial information into spatial knowledge.

Past

> The history of maps is older than history itself, if we think of history
> as beginning with written records.
>
> (Raisz, 1948)

The initial spark

Before the latter half of this century cartographers focused most of their
attention on the production of maps (McCleary, 1970). Maps were judged
on their artistic merits as objects to be appreciated. The ability of the map
to communicate information and the cognitive processes used by map readers
were not a major consideration. A few early studies that considered how
humans interacted with maps can be found (Gulliver, 1908; Trowbridge,
1913); and cartographers showed some concern with map readers' abilities
to compare volumes on population maps (Smith, 1928) or process the
intended visual effect with dot maps (Birch, 1931).

Arthur Robinson, however, is usually credited as the first to link map
design to human cognitive abilities (Robinson, 1952; MacEachren, 1995).
Robinson's goal for cartography was to provide a scientific basis for map
design. He asserted that *function* should provide the basis for design. He
argued for the 'development of design principles based on objective visual
tests, experience, and logic' (Robinson, 1952: 13). He further pointed out
that 'research in the physiological and psychological effects of colour; and
investigations in perceptibility and readability in typography are being
carried out in other fields' and that 'cartography cannot continue to ignore
these developments' (Robinson, 1952: 13–14).

Robinson's (1967) later essay on the psychological effects of colour in
cartography gave the makers of maps much to consider about the role of
colour in graphical communication. He argued that colour is a great simplifying and clarifying element. It can be used as a unifying agent as when
it defines a figure–ground relationship. People have an emotional reaction
to colours that is separate from any rational thinking. He suggested that
'everyone, upon looking at a coloured map, usually instinctively likes it or
dislikes it, quite apart from any understanding of what the map is supposed
to be communicating' (Robinson, 1967: 53). He also argued that colour
can effect the 'general perceptibility' of a map by affecting the legibility
of the lettering and enhancing the map-reader's ability to distinguish fine
details. He went on to discuss a number of problems colour creates for
cartographers and map readers. For example, people are more sensitive and
react faster to some colours than they do to others. Some colours tend to
advance (red) and others retreat (blue) relative to the background. The
simultaneous contrast effect causes hues to appear different when they are
adjacent to the way they appear when they are separated.

Power functions

Some of the earliest efforts to follow Robinson's directions adopted methods used in psychophysics that considered the relationship between a stimulus and the sensed effect of the stimulus (Stevens, 1975). Using regression techniques that were becoming popular in geography, cartographers could fit a power function that connected a relevant stimulus and response (Williams, 1956; Flannery, 1956). The independent variables objectively measured characteristics of map symbols. The dependent variables were judgements of the characteristics made by human subjects.

Ekman *et al.* (1961) had subjects judge the volume of perspective drawings of cubes and spheres. Although power functions could be fit for the data, it was concluded that the subject's responses reflected the perceived areas of the symbol rather than the intended volumes. This clearly indicated that the best intentions of the cartographer might be misdirected if they are based on false assumptions regarding the map reader's cognitive abilities.

Meihoefer (1969) expressed a concern about traditional methods used to construct graduated symbol maps. He argued maps with symbols that were not visually distinguishable were difficult to interpret. He also indicated that 'subjects had unwarranted confidence in their ability to perceive the different size circles', but 'only a small per centage of the subjects tested could distinguish subtle increases in circle size' (Meihoefer, 1969: 115). Additional experiments lead him to conclude that range-graded circles should always be used (Meihoefer, 1973).

Kimmerling (1975) had subjects select grey circles to partition a scale from white to black in equal steps. He then compared the functional relationship between mean estimates of grey experienced by the subjects and percentage reflectance for equally spaced grey values between 0 and 100 per cent. He reported significant differences when estimates were done with a white and black background. He also suggested that the method used to acquire estimates from subjects could influence the results and that performing the same experiment in a map context would be difficult because simultaneous contrast effects could not be held constant.

About the time Kimmerling's paper appeared critics began to point out limitations of the psychophysical studies. Petchenik (1975) pointed out the paradigm shift that was taking place in psychology away from strict behaviourism toward what is now called cognitive psychology. The early power function studies, since they were limited to computing a functional relationship between a stimulus and a response, were clearly not studying cognitive processes used by map readers to acquire symbolic information from the maps and determine the meaning of the symbols. These studies provided cartographers with no theoretical understanding of the interactions taking place between a map and a map reader.

Cox (1976) soon pointed out that the reference symbol used to make judgements of circle sizes had a significant effect on those judgements.

If a small symbol was used in the legend then underestimation occurred and if a large symbol was used in the legend then overestimation occurred. Psychological studies of the *symbolic distance effect* and *semantic congruity effect* provided a general theoretical explanation of this phenomenon (Holyoak and Mah, 1982; Kerst and Howard, 1978). Using reaction times to determine how quickly objects could be compared, it was consistently shown that reaction times decrease with the separation of objects on a scale such as size. A large and a small symbol can be compared faster than two small symbols or two large symbols. Reaction times for the question of which is larger or which is smaller depend on the relative sizes of the pair of objects in question. When the objects in question are relatively large (an aeroplane and a truck), deciding which is larger produces faster times. When the objects are relatively small (moth or fly) deciding which is smaller produces slower times (Holyoak and Mah, 1982).

Chang (1977) argued that the methods used to estimate the sizes of circles affected the estimated parameters of the power functions relating these estimates to actual sizes. Underestimation of circle sizes was found when the ratio of the size of a circle and a referent was estimated but was only a special case when the direct magnitude of a circle was estimated with a given a referent. Chang (1980) also suggested the instructions given the subjects affected the power function parameters. Different exponents were computed when studies requested subjects to judge the *area* of symbols or the *size* of symbols. He also reported the range in symbol size and the order in which the size estimates were obtained also influenced the exponent.

Castner (1983) reviewed the literature on cartographic design related to map reading and communication. He expressed a number of concerns about translating the results of map reading studies into valid map design practices. For example, problems can occur if we focus on just one part of a complex process. If human error is inevitable, should cartographers follow Meihoeffer's advice and construct graduated symbol maps based on ranges of size rather than individual sizes? Is this not 'exchanging the error inherent in human perceptions for one imbedded in the statistical classes themselves'? (Castner 1983, 95).

The experimenter must allow some factors in the experiment to vary and control others. This causes simplified displays to be used in most map reading experiments. The results of experiments on simplified maps tell us something about the map readers' responses to the experimental stimuli, but are these findings directly applicable to designing a complex map? Although this is a valid concern, the answer is clearly not doing uncontrolled experiments with complex maps. Studying cognitive processes used in map reading under controlled conditions reflects the general knowledge and abilities of map readers. The simple models developed from such studies allow a general understanding of the interaction between a typical map reader and map. Studying complex maps that are more realistic, but have many specific characteristics that are not controlled in the experiment, may only indicate how subjects responded to a specific set of maps. This is less likely to produce results that can be generalized beyond the specific maps studied.

Gilmartin (1981, p. 9) argued 'there is a need for both psychophysical and cognitive research in cartography'. These needs are not always compatible and can divide the discipline rather than unite it. Medyckyj-Scott and Board's (1991) essay on cognitive cartography explained these dual needs. They start with Long and Dowell's (1989) assertion that a discipline whose knowledge is based on the practice of its craft often fails to be fully effective. Cartography is a craft in the sense that 'it solves the problems of map design by carrying out design and evaluation, the latter generally informally' (Medyckyj-Scott and Board, 1991: 202).

The process of map making is supported by knowledge that is, for the most part, implicit and informal. Experience produces map construction conventions that are in the form of rules and heuristics. The heuristics are generated for specific problems and tested for effectiveness. Effective map construction procedures become conventions and ineffective procedures are discarded. The testing of individual maps as effective communication devices is generally not possible so the effectiveness of the conventions of map making can not be guaranteed. Since craft knowledge is difficult to test, it is also difficult to generalize. A general theory of map design, therefore, has not been possible. Each map produces a new design problem and is solved by the craftsman using conventions and newly invented heuristics.

Cartographers practising the craft were seeking from the psychophysical studies a simple heuristic, a rule, a tool, a simple power function that could be used to make maps. The focus of psychophysical studies was clearly on the making of the map rather than an understanding of the cognitive processes used in map reading. The focus of cognitive studies in map reading is clearly on how map readers process and determine the meaning of information acquired from maps. The question 'how can I use the results of your cognitive study to make a better map' is frequently asked by map makers, but can seldom be answered in a simple way by cognitive scientists. The goals of the craftsman and the goals of the cognitive scientist are not the same. The map stands in the middle waiting to be constructed and used.

Present

> Whatever task is concerned, the initial stage depends on the two processes of *detection* and *discrimination*. At the most basic level, the map user must be able to respond to what is there: that is, the symbols on the map must be a sufficient stimulus to make them detectable.
>
> (Keates 1982)

People generally read maps because they need information. They may want to know *where* a particular place is located in absolute terms, e.g., its latitude and longitude, or in relative terms, e.g., how close it is to a known reference point. A reference map can usually provide this type of information. At other times the map reader wants to know *what* a place is like. Specialized thematic maps provide this type of information, e.g., a graduated symbol map showing population size of cities. These two examples assume the person knows the name of the place of interest. After identifying the name on the map, the absolute or relative location and the population size can be easily determined.

Map reading is an integration and synthesis of knowledge. Even these simple examples require both bottom-up information (the map) and top-down information (information stored in the map reader's memory). The bottom-up information is contained in the lines, colours, shapes, words, etc. the cartographer has put on the map. The top-down information is prior knowledge previously acquired by the map reader (Figure 6.2). It might be general factual knowledge learned in another context, but applicable to map reading, e.g., the meaning of words or the common names for colours. It might be cognitive abilities to process information that was learned and practised in other contexts, but applicable to map reading. Examples might be judging distances between objects, comparing sizes of objects, distinguishing colours, counting, reading, etc. Other top-down information is prior knowledge about maps and the conventions used by cartographers to produce maps. Examples might be the meaning of common symbols such as the North arrow, the graphic scale, and contour lines. Other examples might be the conventional uses of colours and shapes, e.g., blue for water and stars for capital cities. The discussion that follows first considers studies related to perceptual processes used during map reading. Visual search is generally important to any type of map reading. Object files are more particularly related to cartographic animations. The final discussion in this section of the paper considers how the information acquired from maps is stored in our memory. Maps have been shown to be very useful in the creation of mental models.

Perceptual processes

Visual search

The viewer of a typical map should experience objects with information on a number of separable dimensions, e.g., colour, shape, or size (Garner, 1974; Shortridge, 1982; Dobson, 1983). Theories of visual search, such as the *Feature Integration Theory of Attention* and the *Guided Search Theory*, explain how visual information is integrated and how spaces, such as maps, are searched (Duncan and Humphrey, 1989; Cave and Wolfe, 1990; Chun and Wolfe, 1996; Wolfe *et al.*, 1989; Treisman, 1988; Treisman and Gelade, 1980;

Figure 6.2 Map reading is an integration and synthesis of bottom-up and top-down information.

Wolfe, 1994). These theories consider how information is acquired from a visual display during an initial parallel stage of a search process and used in a subsequent serial stage to identify the target of the search. Objects located on a cartographic map would activate feature modules in memory during the parallel stage. The activation for particular features, e.g., colour, size, or shape, is dependent on both bottom-up information (in the display) and top-down information (in the viewer's memory). Targets with unique features can be identified early in the process and are said to 'pop out' of the display. Targets that share features with other objects on the map are more difficult to find and usually require a serial search before they can be identified.

A number of visual search studies have been conducted with maps that have focused on map symbols (Lloyd, 1988, 1997a), locations with particular colours (Nelson, 1994), and boundaries with particular colours (Brennan and Lloyd, 1993; Bunch, 1998). Lloyd (1988) investigated the search for pairs of pictographic map symbols on cartographic and cognitive maps. His objective was to determine if searching for symbols previously encoded in memory was the same as searching for symbols currently being viewed on a map. Perception subjects were first shown a pair of yellow pictographic symbols to remember. Maps were presented with a green island surrounded by a blue background and a number of symbols. The maps included the target symbols on half the trials and each map had from four to twenty-four symbols. Subjects simply indicated if both symbols were on a trial map (Yes or No). As expected, results for the perception subjects indicated they were using a serial self-terminating process to perform the search task.

Memory subjects followed a similar procedure except for the order of presentation. Memory subjects were first shown a map for a given trial that had a fixed number of symbols and they encoded the symbols into memory. Two symbols were then presented and the subject determined if they were on the previously viewed map (Yes or No). Reaction times for the memory subjects were not significantly related to the number of symbols on the map. This indicated they were using a parallel search process. It was concluded that search processes used by memory and perception were not functionally equivalent.

Lloyd (1997a) considered variations in visual search for targets on maps and focused on differences between searching for targets with unique features and targets that share features with other objects. Experimental map symbols were constructed using colour, shape, size, and orientation. The study considered the relative importance of these dimensions as well as the location of the target on the map. The efficiency of spatial search was considered the time needed to find a target in a display. The best 'pop out' effect occurred when colour was used to produce unique differences between targets and other symbols. Targets defined by orientation differences produced the slowest search times. A 'pop out' effect also occurred when colour was combined with shape to define the target even when the target shared features with other symbols. The location of the target on the map also had a significant effect on reaction times. Targets in the centre of the map were found faster because most subjects starting searching in the centre of the display.

Nelson (1994) has conducted experiments that considered searching for colour targets on choropleth maps. Results additionally demonstrated that differences between the target and distractors were the most important factor affecting the search process. Nelson also reported that it was difficult to create maps that represented Duncan and Humphrey's (1989) hypothetically most difficult search condition. This extreme case, with the target being very similar to the distractors, but the distractors being very dissimilar to each other, may be impossible to produce on real maps.

In a study conducted by Brennan and Lloyd (1993) subjects searched choropleth maps for a target boundary among other boundaries. Boundaries were defined by two polygons with two specific colours on either side of the boundary. Reaction times and the accuracy were analysed to test hypotheses related to the search process being used by subjects. Their results indicated that both parallel and serial processes were important components of visual search (Cave and Wolfe, 1990). The spatial locations of the target boundaries had a significant effect on the efficiency of the search process. This suggested subjects were serially processing the locations on the map in a systematic fashion and that targets in some parts of the map were easier to detect. As the number of boundaries with one of the target colours (critical boundary) increased, reaction times also tended to increase. As the number of boundaries without either boundary colour (non-critical boundary) increased reaction times tended to decrease. Maps with few critical boundaries and many non-critical boundaries had the fastest response times. This result suggested distractors with common characteristics can be suppressed as a category rather than individually to save processing time (Duncan and Humphrey, 1989). Some colour combinations used to define boundaries produced faster response times than other combinations. Opponent process theory was used to explain these results. The receptors responsible for detecting a particular opponent pair, i.e., red or green, blue or yellow, and black or white, can only detect one of the two colours at any point in time. Boundaries defined by two colours that were not from the same opponent pair, e.g., red and blue, should be detected faster than boundaries defined by colours that were from the same opponent pair, e.g., red and green or blue and yellow.

Bunch (1998) performed a similar study using boundaries defined by opponent colours that also varied in both hue and luminance. Colours for the study were created in a CIELab colour space so characteristics on red-green, yellow-blue, and luminance axes could be controlled (Abramov and Gordon, 1994). For example, the yellow colours had values that produced perceptually unique yellow on the yellow-blue axes and zero on the red-green axis. Results generally indicated that faster times occurred when the boundary colours were separated in the CIELab space. Faster times were also recorded for boundary colours that had the same hue but differed in luminance. This combination created a target with a unique hue that would 'pop out' of the map.

Object files

Those interested in producing animated maps should be interested in object file theory because it explains how map readers theoretically respond to what they see on animated maps. Kahneman and Treisman (1984) were the first to discuss object files as temporary episodic representations of real world objects. They wanted to make a clear distinction between object files as temporary representations of objects and other representations of the objects stored in long-term memory. The more specific and temporary object files were called *tokens*, and the more generalized and permanent structures used to label the object's identity were called *types*. A *token* is a specific exemplar currently being viewed and a *type* is a generalized representation of the category of object in long-term memory.

In a later paper Kahneman *et al.* (1992) argued that focusing attention on a particular object enhanced the salience of its current properties and also reactivates a recent history of the object. The process that causes a current object to evoke an item previewed in a previous visual field was called *reviewing*. As an object moves, changes characteristics, or momentarily disappears from sight the reviewing process maintains the perceived continuity of the object by relating its current state to its previous states. The reviewing process assists recognition of an object when previous and current states of an object match, but hinder it when they do not match.

An object file is opened for use by the perceptual system once we have focused attention on the object. Information is put into the file to record the changes occurring to the object over time. This is necessary to maintain the continuity of a changing object. Kahneman *et al.* (1992) made a distinction between identifying and seeing an object. At first sight you may not be able to identify an object by name. You may see an object on an animated map, open up object file X, and track the object. A name for this X-file may be added later when you have consulted the legend for more information.

While viewing a map animation one may need to determine if there have been two objects on the map or one object that has changed it characteristics. The perceptual system can usually maintain the continuity of a single file and either keep it open or close the first file and create a second file. Our experiences with objects enable us to predict expected motions and behaviours that can be used to make commonsense decisions about objects we are viewing. 'The identity of a changing object is carried by the assignment of information about its successive states to the same temporary file, rather than by its name or by its properties' (Kahneman *et al.*, 1992: 178). An object could change all of its characteristic and its label over time and still be considered the same object under some circumstances. For example, a caterpillar could disappear inside a cocoon and emerge as a butterfly. If an apparent change is detected in an object, we must decide if the current information related to the change, e.g., a new location or new colour, needs to be assigned to the existing object file or a new object file needs to be created.

When multiple objects are in a display, *correspondence* must be determined before apparent motion can be experienced. The correspondence operation determines if each object in the second view of the display is a new object or an old object in a new location. Geographers should find it interesting that spatial and temporal constraints rather than other characteristics, e.g., colour, size, or labels, primarily control this operation. Apparent motion is only perceived when spatial and temporal proximity is maintained. Other properties of the object, however, can change without a new object file being created. *Reviewing* of the object file must be completed to retrieve the characteristics for the object in the initial display that are no longer visible. Apparent motion occurs because this earlier information provides the context for the change. An *impletion* process uses the reviewed information from the first display and the current information to produce the perception of motion that links these displays.

Cammack (1995) investigated the role of object files in map reading. His results indicated no consistent reaction time advantage when objects on target maps were related to objects on preview maps by shape or colour. This supported the notion that changes in characteristics other than location are not of primary importance in maintaining object files. The results of a second experiment indicated objects moving shorter distances could be named significantly faster than objects moving longer distances. Objects that moved rapidly could also be named significantly faster than slower moving objects. This supported arguments that spatial and temporal proximity must be maintained for the apparent motion of a single object to be perceived.

Memory processes

Mental models

Once we have acquired information from a map we need to integrate it with information acquired from other sources and store it in some organized structure in our memory. A mental model is a cognitive structure that has been encoded to represent a person's spatial knowledge of an environment. A person's mental model of an environment represents that person's learned knowledge of the environment at a point in time that can be updated as new information is acquired from additional experiences with the environment. Studies have shown that mental models can be encoded by viewing maps or navigating in an environment (McNamara *et al.*, 1992; Taylor and Tversky, 1992a, 1992b) or by reading verbal descriptions (Bryant *et al.*, 1992; Byrne and Johnson-Laird, 1989; Ferguson and Hegarty, 1994; Franklin *et al.*, 1992; Taylor and Tversky, 1992a, 1992b). 'Human beings can evidently construct mental models by acts of imagination and can relate propositions to such models' (Johnson-Laird, 1983: 156).

Taylor and Tversky (1992b) suggested that mental models are like an architect's three-dimensional model of a city that can not be viewed as a finite entity. Johnson-Laird (1983) argued that images are created to provide particular views of the model from specific perspectives, but the model cannot be visualized as a whole. Our ability to visualize environments from multiple perspectives supports the contention that images can be created from mental models (Bryant *et al.*, 1992; Franklin *et al.*, 1992).

Researchers have suggested that mental models integrate spatial and verbal information into a single structure (Glenberg and McDaniel, 1992; McNamara *et al.*, 1992). It is also thought that comprehension is increased if mental models are created while a text is being read (Glenberg and Langston 1992; Waddill and McDaniel 1992; Wilson *et al.*, 1993). It has been reported that a text comprehension was better for subjects studying route texts over survey texts (Perrrig and Kintsch, 1985) and that subjects who read route texts constructed more accurate sketch maps than subjects who read survey texts (Ferguson and Hegerty, 1994). This latter effect was eliminated when a map accompanied the texts.

The flexibility of mental models makes them effective representations of spatial environments. They can be used for a variety of tasks because the models integrate visual and verbal information. They could be used to plan the route between two locations in an environment (Taylor and Tversky, 1992a) or to create multiple images to view an environment from different perspectives (Franklin *et al.*, 1992).

Taylor and Tversky (1992b) argued that mental models of environments do not have a single perspective. This argument was based on results that indicated subjects learning an environment from route or survey descriptions were equally fast and accurate verifying inference statements presented in either perspective. The authors argued that subjects used a mental model to verify inference statements because they could not compare inference questions directly to a representation of the text. Since inference statements were responded to with equal efficiency when a statement's perspective did not correspond to the perspective represented by the text, it was argued that mental models have no perspective. It was also reported that subjects who studied maps of the environments also created mental models with no perspective.

Lloyd *et al.* (1995) have shown that perspective free mental models of a city can be encoded by reading both route and survey description of how to go from one landmark to another, by seeing the routes traced on a vertical perspective map, or by watching video simulations of driving the routes along streets. Although all subjects were able to encode the spatial information about the city and produce an accurate sketch map, the methods used to encode that information did produce some significant differences in the nature of the mental models.

Subjects who read survey text descriptions were more accurate than subjects who read route descriptions. The two groups had response times that were not significantly different. Subjects who viewed maps to encode the information had significantly slower response times than subjects who viewed videos of driving the routes. Both of these groups were equally accurate. It also was reported that those who viewed the video took much more time to encode the information.

Future

> The real problem is this: How does a map user develop internal, personal knowledge of relations among things in space on the basis of viewing a sheet of paper covered with ink marks? How, in common language, does one read a map?
>
> (Petchenik, 1975)

Predicting the future of cartographic research is not a simple task. It should be safe to say that understanding the cognitive processes used by map readers will continue to be an important goal for cartography. As those interested in spatial cognition advance toward this goal, those who practice the craft of making maps should benefit from the insights provided by cognitive research. This section of the paper discusses one broad issue related to a research perspective that could have a significant impact on cartographic research (connectionist models) and another, more specific, issue related to learning with maps (prior knowledge).

Connectionist theory

If cartographers are to develop a theory of map reading, they need to develop a perspective that allows them to (1) observe processes that directly connect map readers with their maps (2) form hypotheses that explain these processes, and (3) develop theories that connect these processes and ultimately provide an explanation of how people read maps. Two major obstacles have made it difficult to develop a theory of map reading. First, the cognitive processes are not easily observed. They take place in the brain of a map reader who, for the most part, is unaware of how he is processing information. Second, most map reading tasks are complex and involve multiple cognitive processes that need to be understood.

Models such as the cartographic communication model (Figure 6.1) are very general expressions of the elements involved, but do not tell us what we need to know. Cartographers have generally relied on information from laboratory experiments that provided surrogate measures of activities going on in the brain such as reaction times and accuracy to test hypotheses on map reading. Although these 'black-box' studies have been very productive, cognitive scientists from a variety of disciplines have argued that models based on biologically correct principles have a better chance of evolving into credible theories (Kohonen, 1989; Kosko, 1993; Kosslyn and Koenig, 1992; Lloyd 1997b; Martindale 1991).

Such theories have been referred to as connectionist theories because they are based on models of how neurons connected as a network in the brain process information. Such studies frequently use artificial neural networks which can learn the same information that humans are learning with their biological equivalents (Hewitson and Crane, 1994). Insights into human learning may be possible by comparing human behaviour to neural network simulations. The patterns of connection weights between neurons in artificial neural networks can provide previously unavailable information for interpreting how people process information to complete a task (Lloyd and Carbone, 1995).

Lloyd (1994) conducted a map learning experiment that had both human subjects and *auto-associative* neural networks learn spatial prototypes. This type of neural network is capable of learning a prototype from a series of input patterns (McClellend and Rumelhart, 1985). The prototype is the generalized most typical pattern for the category. The prototypes were for categories of maps that were supposed to be made by a number of European countries during the age of discovery. Critical information on the map related to the location, size and orientation of an island on each of the maps. An analysis based on typicality ratings indicated the human subjects and the neural networks had learned similar prototypes for the categories. An inspection of the connection weights in the neural networks indicated how a prototype was learned and stored in memory for each category. Another interesting finding was that the neural networks did a better job in learning some prototypes because they were not as influenced as humans were by the order in which they experienced the individual maps.

Cammack and Lloyd (1993) constructed an *Interaction Activation Competition* neural network that simulated the storage and retrieval of information from associative memory (McClellend and Rumelhart, 1989). The idea was to produce a memory structure that had learned accurate census information about the states of the US. The information encoded into the artificial cognitive map related to population, employment, race, politics, income, and regional location. By initially activating a particular state in the network one could spread activation throughout the network and turn on neurons that represented the characteristics of the particular state and any other states with similar characteristics. By initially activating a particular neuron representing a characteristic, e.g., high income, one could spread activation throughout the network to activate states with this characteristic or other characteristics, e.g., voting Republican, associated with this characteristic. If a collection of states representing a region, e.g., the South, were initially activated, then activation would spread throughout the network to activate generalized characteristics of the region. This type of network can simulate the storage and retrieval of information in a cognitive map related to *where* places are and *what* they are like. By encoding a similar model for human subjects who had recently learned the same information presented on maps one could use the *Interaction Activation Competition* model to make inferences about what was actually learned during the map reading experience.

Lloyd (1998) simulated the learning of an outline map of Texas using a Kohonen neural network (Kohonen, 1989, 1993). A Kohonen network consists of neurons in an input layer that provide the information to be learned. These neurons are connected to another layer of neurons that is usually called the Kohonen self-organizing map. The model is called self-organizing because it does not learning by correcting activation errors. Instead, it learns by recognizing patterns in the information being presented to it without the benefit of comparing its output to the truth.

In this case, physical space coordinates for boundary points along the outline of Texas were presented to the network one point at a time. The neurons in the Kohonen self-organizing map were connected to the input neurons by weights that were initialized to small random numbers to begin the process. For this type of model the coordinate information for each point is multiplied by the current weights and summed to produce an activation value for each neuron in the Kohonen layer. The neuron with the highest activation is declared the winner. Only this neuron and its neighbours are allowed to change their weights to be more like the input data. After cycling the points through the network many times the input patterns (physical coordinates) consistently activate the same winning neurons in the Kohonen layer. When this happens the coordinates from the cartographic map have been projected into the Kohonen self-organized map. By fixing the weights of the network and passing the boundary points through again one can compute the standardized coordinate location of each point in the Kohonen self-organizing map. A comparison of the cartographic map of Texas (Figure 6.3a) and the aggregate map learning by thirty Kohonen neural network simulations (Figure 6.3b) indicated the typical simplified shape learned by the neural network. Basic sections of the boundary can easily be identified on the distorted representation. It is difficult to judge how well the Kohonen network models the process used by humans to acquire boundary information for Texas. A comparison can be made with the representation of Texas aggregated from fifteen geography graduate students' sketch maps (Figure 6.3c). The students drew the boundaries from memory without any particular training to know the location of the forty-two boundary points learned by the neural network. The even more simplified aggregate map recalled by the students may be partially due to their inability to express their knowledge through drawings or because they never actually encoded the details of the Texas boundary.

Learning and prior knowledge

Understanding how spatial knowledge is learned from maps and stored in long-term memory is a challenging research topic. It would seem reasonable to assume that things we already know might help us learn new things more effectively. Someone who has practised reading maps and who has become familiar with the conventions used by cartographers to make maps should be able to learn more effectively from a map than an inexperienced and uninformed person.

Dochy's (1991) review of the literature suggested the term 'prior knowledge' has been defined and used many different ways by researchers. In a

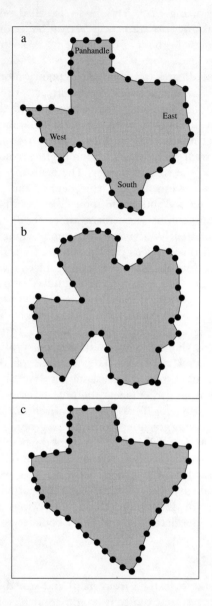

Figure 6.3 (a) The forty-two points from the cartographic map of Texas; (b) learned by the Kohenen neural network and represented in the Kohenen layer as a self-organized cognitive map; (c) the same forty-two points aggregated from sketch maps made by fifteen geography graduate students.

map reading context it might refer to declarative knowledge such as the meaning of symbols on a map or the convention that certain colours are consistently used to symbolize types of objects on maps. It might also refer to procedural knowledge such as how to estimate distances between objects or how to determine the elevation of a location. At times episodic knowledge may be important if one needs to recall information from a 1950 map to compare it with information from a 1990 map. Semantic knowledge is also important in that one must know what the category *lake* or the category *dune* means if you read it on a map. Finally strategic knowledge can make one's map reading experience more efficient. For example, the time it takes to search for the highest point on a map might be reduced by a plan that included identifying the direction that a stream flows and following it upstream to higher elevations.

Dochy (1988) suggested a number of theoretical reasons why prior knowledge would be an important aid for learning. Map readers with previous related experiences might be able to structure the new information in long term memory more efficiently than map readers without such experiences. The development of connections between newly learned information and previously learned information can result in multiple and redundant retrieval paths in the cognitive structure storing the information. Access to prior knowledge during the learning of new information can reduce the load on working memory and result in more information being processed in a fixed amount of time. Prior knowledge may be used to direct attention in a more selective way so that more useful information is processed and less useful information is ignored. If prior knowledge increases the total amount of information available during the learning process, then more total information is likely to be retained. The activation of prior knowledge on a specific topic generally aids the retrieval of information from memory. Prior knowledge might be in the form of an organized knowledge structure that guides perception (schema) and aids in the storage and retrieval of information. Prior knowledge saves encoding time because fewer propositions about the new information need to be encoded if they already exist.

Three studies of learning by reading maps are discussed below. The first two involve children learning from map reading experiences and the third shows that prior knowledge may have both positive and negative consequences.

Animation

Animated maps have been considered as a new improved approach to map communication, but studies of this potential have been inconsistent (Patton and Cammack, 1996; Slocum and Egbert, 1993). Cutler (1998: 20) considered whether 'children would be able to comprehend and recall more information from an animated isoline map than from traditional paper isoline maps'. Some seventh-grade children viewed either an animation of the decline of cotton in South Carolina (animation group) depicted as isolines that appeared, changed locations, and disappeared during a fifty year period. Other seventh-grade children viewed the same decline of cotton in South Carolina depicted on a sequence of six paper maps (map group). After the learning experience was completed students were tested on the information they had experienced relating to the decline of cotton in South Carolina.

Results indicated that a subject's prior knowledge of isoline maps and reading level significantly affected both the times it took to answer questions and the accuracy of the answers. Students who scored higher on a previous test on interpreting temperature isoline maps were found to be both faster and more accurate on the decline of cotton test. Students who were categorized as reading at a higher level by their teacher were also faster and more accurate than other students. Although students in the animation group took slightly longer to respond to the test questions than students in the map group, there was no significant difference in reaction times. The students learning from paper maps, however, were significantly more accurate than the students who viewed the animation. It would appear that the characteristics of the map readers were more important in predicting successful learning than the characteristics of the map presentations. A significant interaction effect between reading level of the students and type of presentation produced a very interesting result. Students who were reading at a higher level were able to respond faster if they had experienced the paper maps, but students who were reading at a lower level responded faster if they had experienced the map animation. This suggested that students who were reading at a lower level could gain some advantage from viewing animations. Learning from paper maps consistently resulted in more accurate test scores for students at all reading levels.

Cartograms

Breding (1998) had some seventh-grade students study maps and learn the information presented on them. Some students studied maps of a familiar home region and others studied an unfamiliar region. The maps were constructed in two ways. Some had graduated circles showing population sizes of individual locations and others were cartograms also showing population sizes of individual locations. It was assumed that students who were using maps of a familiar region with familiar names and students who were using maps with commonly used familiar symbols would have some useful prior knowledge and be advantaged in their learning; equally, students learning from maps of an unfamiliar region with unfamiliar names and unfamiliar cartograms would be disadvantaged in their learning. After learning the population size information on the maps, students were given a test to assess what they had learned.

Results indicated that familiarity with the regions significantly affected response times but did not significantly affect accuracy. It would appear that prior knowledge of the familiar regions helped students store the information in a more accessible structure, but this had no effect on the accuracy of the learned information. Students who studied familiar graduated circle maps were able to respond faster to the test questions, but they had the lowest accuracy. One explanation of this was that the simple shape of circles made it difficult to remember which circle was in what location. Since the cartograms showed population sizes with unique shapes this type of confusion is less likely. Familiarity may simplify learning, but working harder to acquire the information may pay off with a higher accuracy or a longer retention.

Patton (1997) considered the effect of prior knowledge on the learning of category information from maps. Subjects learned to categorize maps that appeared on a computer monitor into either category *A* or category *B*. A coherent condition presented *A* and *B* categories that were related concepts such as *Urban* and *Rural*. These maps would have symbols associated with the categories. One of the symbols on an *Urban* maps might be a high-rise apartment building and a symbol on a *Rural* map might be a barn. An incoherent condition presented *A* and *B* categories that had symbols with no apparent pattern. Symbols on the maps were sometimes relevant to the category (a tractor on a *Rural* map) and sometimes irrelevant to the category (a flag on a *Rural* map).

When the categories were coherent the information was learned significantly faster than when the categories were incoherent. This expected result suggested that the prior knowledge about categories such as *Urban* and *Rural* made it easier to learn which map belonged in which category. When subjects were required to identify maps as *Old* maps that they had learned or *New* maps they had not seen before, they were more accurate at identifying *Old* maps if they had learned coherent categories and less accurate with incoherent categories. There was also evidence that prior knowledge caused more frequent errors when considering *New* maps. Patton (1997) called one of the new maps judged the prototype map because it contained symbols that were frequently on the learned maps and were associated with the category. The subjects, however, had never seen this map. Subjects who learned coherent categories wrongly identified the prototype more often as an *Old* map than subjects who learned incoherent categories. The prior knowledge made it much easier to learn the maps for coherent categories, but they apparently did not know as many specific details as subjects who learned the maps for incoherent categories.

A unified theory of map reading may not easily be written because map reading involves many complex processes. The path to that theory, however, will undoubtedly involve studies on how visual information is processed, integrated with prior knowledge, and used to make judgements and decisions.

References

Abramov, I. and Gordon, J. (1994) Color appearance: On seeing red – or yellow, or green, or blue. *Annual Review of Psychology*, 45, 451–85.

Birch, T. (1931) *Maps: Topographical and Statistical*. Oxford: Clarendon Press.

Board, C. (1967) Maps as models. In Chorley, R. and Haggett, P. (eds), *Models in Geography*. London: Methuen, pp. 671–725.

Breding, P. (1998) The effect of prior knowledge on eighth graders' abilities to understand cartograms. Unpublished MA thesis, Department of Geography, University of South Carolina.

Brennan, N. and Lloyd, R. (1993) Searching for boundaries on maps: The cognitive process. *Cartography and Geographic Information Systems*, 20, 222–36.

Bryant, D., Tversky, B. and N. Franklin (1992) Internal and external spatial frameworks for representing described scenes. *Journal of Memory and Language*, 31, 74–98.

Bunch, R. (1998) Searching for bi-polar color boundaries on maps. Unpublished MA thesis, Department of Geography, Univeristy of South Carolina.

Byrne, R. and Johnson-Laird, P. (1989) Spatial reasoning. *Journal of Memory and Language*, 28, 564–75.

Cammack, R. (1995) The cognition of cartographic animation with object files. Unpublished PhD dissertation, Department of Geography, University of South Carolina.

—— and Lloyd, R. (1993) Connected space: Regional neural networks. Paper presented at the Annual Meeting of the Association of American Geographers, Atlanta, Georgia.

Castner, H. (1983) Research questions and cartographic design. In Taylor, D. (ed.), *Graphic Communication and Design in Contemporary Cartography*. Chichester: Wiley, pp. 87–114.

Cave, K. and Wolfe, J. (1990) Modeling the role of parallel processing in visual search. *Cognitive Psychology*, 22, 225–71.

Chang, K. (1977) Visual estimation of graduated circles. *The Canadian Cartographer*, 14, 130–8.

—— (1980) Circle size judgements and map design. *The American Cartographer*, 7, 155–62.

Chun, M. and Wolfe, J. (1996) Just say no: How are visual searches terminated when there is no target present? *Cognitive Psychology*, 30, 39–78.

Cohen, A. and Rafael, R. (1991) Attention and feature integration: Illusory conjunctions in a patient with a parietal lobe lesion. *Psychological Science*, 2, 106–10.

Cox, C. (1976) Anchor effects and the estimation of graduated circles and squares. *The American Cartographer*, 3, 65–74.

Cutler, M. (1998) The effects of prior knowledge on children's abilities to read static and animated maps. Unpublished MA thesis, Department of Geography, University of South Carolina.

Dobson, M. (1983) Visual information processing and cartographic communication: The utility of redundant stimulus dimensions. In Taylor, D. (ed.), *Graphic Communication and Design in Contemporary Cartography*. New York: Wiley.

Dochy, F. (1988) The prior knowledge state of students and its facilitating effect on learning. *OTIC Research Report* 1–2.

—— (1991) Mapping 'prior knowledge' or 'expertise': A tentative outline. *OTIC Research Report* 28.

Duncan, J. and Humphrey, G. (1989) Visual search and stimulus similarity. *Psychological Review*, 96, 433–58.

Ekman, G., Lindman, R. and William-Olsson, W. (1961) A pyschophysical study of cartographic symbols. *Perceptual and Motor Skills*, 13, 355–68.

Ferguson, E. and Hegarty, M. (1994) Properties of cognitive maps constructed from texts, *Memory and Cognition*, 22, 455–73.

Flannery, J. (1956) The graduated circle: A description, analysis, and evaluation of a quantitative map symbol. Unpublished PhD thesis, University of Wisconsin, Madison, WI.

Franklin, N., Tversky, B. and Coon, V. (1992) Switching points of view in spatial mental models, *Memory and Cognition*, 20, 507–18.

Garner, W. (1974) *The Processing of Information and Structure*. Potomac, MD: Erlbaum.

Gilmartin, P. (1981) The interface of cognitive and psychological research in carto-graphy. *Cartographica*, 18, 9–20.

Glenberg, A. and Langston, W. (1992) Comprehension of illustrated text: Pictures help to build mental models. *Journal of Memory and Language*, 31, 127–51.

—— and McDaniel, M. (1992) Mental models, picture, and text: Integration of spatial and verbal information. *Memory and Cognition*, 20, 458–60.

Gulliver, F. (1908) Orientation of maps. *Journal of Geography*, 7, 55–8.

Hewiston, B. and Crane, R. (1994) *Neural Nets: Applications in Geography*. Dordrecht: Kluwer.

Holyoak, K. and Mah, W. (1982) Cognitive reference points in judgments of symbolic magnitude. *Cognitive Psychology*, 14, 328–52.

Johnson-Laird, P. 1983. *Mental Models: Towards a Cognitive Science of Language, Inference, and Consciousness*. Cambridge: Cambridge University Press.

Kahneman, D. and Treisman, A. (1984) Changing views of attention and auto-maticity. In Parasuraman, R. and Davies, D. (eds), *Varieties of Attention*. New York: Academic, pp. 29–61.

—— , —— , and Gibbs, B. (1992) The reviewing of object files: object-specific integration of information. *Cognitive Psychology*, 24, 175–219.

Keates, J. (1982) *Understanding Maps*. New York: Halsted.

Kerst, S. and Howard, J. (1978) Memory psychophysics for visual area and length. *Memory and Cognition*, 6, 327–35.

Kimmerling, J. (1975) A cartographic study of equal value gray scales for use with screened gray areas. *The American Cartographer*, 2, 119–27.

Kohonen, T. (1989) *Self-Organization and Associative Memory*. Berlin: Springer-Verlag.

—— (1993) Physiological interpretation of the self-organizing map algorithm. *Neural Networks*, 6, 895–905.

Koláčny, A. (1969) Cartographic information – a fundamental concept and term in modern cartography, *Cartographic Journal*, 6, 47–9.

Kosko, B. (1993) *Fuzzy Thinking: The New Science of Fuzzy Logic*. New York: Hyperion.

Kosslyn, S. and Koenig, O. (1992) *Wet Mind: The New Cognitive Neuroscience*. New York: Free Press.

Landau, B. (1986) Early map use as an unlearned ability. *Cognition*, 22, pp. 201–23.

Long, J. and Dowell, J. (1989) Conceptions of the discipline of HCI: Craft, applied science and engineering. In Sutcliffe, A. and Macaulay, L. (eds), *People and Computers V*. Cambridge: Cambridge University Press, pp. 9–32.

Lloyd, R. (1988) Searching for map symbols: The cognitive process. *The American Cartographer*, 15, 363–78.

—— (1994) Learning spatial prototypes. *Annals of the Association of American Geographers*, 84, 418–40.

—— (1997a) Visual search processes used in map reading. *Cartographica*, 34, 11–32.

—— (1997b) *Spatial Cognition: Geographic Environments*. Dordrecht: Kluwer.

—— (1998) Learning geographic shapes. Unpublished paper, Department of Geography, University of South Carolina.

—— Cammack, R., and Holliday, W. (1995) Learning environments and switching perspectives. *Cartographica*, 32, 5–17.

—— and Carbone, G. (1995) Comparing human and neural network learning of climate categories. *Professional Geographer*, 47, 237–50.

McCleary, G. (1970) Beyond simple psychophysics: Approaches to the understanding of map perception. *Proceedings of the American Congress of Surveying and Mapping*, 189–209.

McClelland, J. and Rumelhart, D. (1989) *Explorations in Parallel Distributed Processing: A Handbook of Models, Programs, and Exercises*, Cambridge, MA: MIT.

MacEachren, A. (1995) *How Maps Work: Representation, Visualization, and Design*. New York: Guilford.

McNamara, T., Halpin, J. and Hardy, J. (1992) The representaiton and integration in memory of spatial and nonspatial information. *Memory and Cognition*, 20, 519–32.

Martindale, C. (1991) *Cognitive Psychology: A Neural-network Approach*. Pacific Grove, CA: Brooks/Cole.

Medyckyj-Scott, D. and Board, C. (1991) Cognitive cartography: A new heart for a lost soul. In Mueller, J. (ed.), *Advances in Cartography*. London: Elsevier.

Meihoeffer, H. (1969) The utility of the circle as an effective cartographic symbol. *The Canadian Cartographer*, 6, 105–117.

—— (1973) The visual perception of the circle in thematic maps/ experimental results. *The Canadian Cartographer*, 10, 63–84.

Nelson, E. (1994) Colour detection on bivariate choropleth maps: The visual search process. *Cartographica*, 31, 33–43.

Patton, D. (1997) The effects of prior knowledge on the learning of categories of maps. *The Professional Geographer*, 49, 126–36.

—— and Cammack, R. (1996) An examination of the effects of task type and map complexity on sequenced and static choropleth maps. In Wood, C. and Keller, P. (eds), *Cartographic Design: Theoretical and Practical Perspectives*. London: Wiley, pp. 237–52.

Perrig, W. and Kintsch, W. (1985) Propositional and situational representations of text. *Journal of Memory and Language*, 24, 503–18.

Petchenik, B. (1975) Cognition in cartography. *Proceedings of the Symposium on Computer Assisted Cartography*, 183–93.

Raisz, E. (1948) *General Cartography*. New York: McGraw-Hill.

Robinson, A. (1952) *The Look of Maps: An Examination of Cartographic Design*. Madison, WI: University of Wisconsin Press.

—— (1967) Psychological aspects of colour in cartography. *The International Yearbook of Cartography*, 7, 50–9.

Shortridge, B. (1982) Stimulus processing models from psychology: Can we use them in cartography. *The American Cartographer*, 9, 69–80.

Slocum, T. and Egbert, S. (1993) Knowledge acquisition from choropleth maps. *Cartography and Geographic Infomation Systems*, 20, 83–95.

Smith, G. (1928) A population map for Ohio for 1920. *Geographical Review*, 18, 422–27.

Stevens, S. (1975) *Psychophysics: Introduction to its Perceptual, Neural, and Social Prospects*. New York: Wiley.

Taylor, H. and Tversky, B. (1992a) Descriptions and depictions of environments. *Memory and Cognition*, 20, 483–96.

—— and —— (1992b) Spatial mental models derived from survey and route descriptions. *Journal of Memory and Language*, 31, 261–92.

Treisman, A. (1988) Features and objects: The fourteenth Bartlett Memorial Lecture. *The Quarterly Journal of Experimental Psychology*, 40A, 201–37.

—— and Gelade, G. (1980) A feature integration theory of attention, *Cognitive Psychology*, 12, 97–136.

Trowbridge, C. (1913) On fundamental methods of orientation and imaginary maps. *Science*, 38: 888–97.

Tversky, B. (1981) Distortions in memory for maps. *Cognitive Psychology*, 13, 407–33.

—— (1992) Distortions in cognitive maps. *Geoforum*, 23, 131–8.

Waddill, P. and McDaniel, M. (1992) Pictorial enhancement of text memory: Limitations imposed by picture type and comprehension skill. *Memory and Cognition*, 20, 472–82.

Williams, R. (1956) *Statistical Symbols for Maps: Their Design and Relative Values*. New Haven, CT: Yale University Map Laboratory Report.

Wilson, S., Rinck, M., McNamara, T., Bower, G. and Morrow, D. (1993) Mental models and narrative comprehension: Some qualifications. *Journal of Memory and Language*, 32, 141–54.

Wolfe, J. (1994) Guided search 2.0: a revised model of visual search. *Psychonomic Bulletin and Review*, 1, 202–38.

—— Cave, K., and Franzel, S. (1989) A modified feature integration model for visual search. *Journal of Experimental Psychology: Human Perception and Performance*, 15: 419–33.

7 Understanding and learning virtual spaces

Patrick Péruch, Florence Gaunet,
Catherine Thinus-Blanc and Jack Loomis

Introduction

During the last several decades, cognitive mapping has emerged as a major area of behavioural research on humans and other species. The research questions fall largely within three distinct topics. First, what information is used by the agent in forming a cognitive map of the environment? Second, what are the functional properties of the cognitive map that gets established in memory? Finally, what are the neural structures involved in cognitive mapping and how do they co-operate in this processing? Two recent technologies, Virtual Environment Technology (VET) and functional brain imaging, are very likely to advance basic research in the field of cognitive mapping, especially if used together. This chapter will re-examine some of the above issues in the light of these new technologies and will try to draw some features for future research in the area of cognitive mapping.

Cognitive maps and cognitive mapping: a brief overview

The term 'cognitive map' was first coined by Tolman (1948), on the basis of data observed in rats, when he tried to explain the processes underlying the spatial behaviour in terms other that of a purely behaviourist (stimulus–response) point of view. Since Tolman, the term 'cognitive map' has come into wide usage and is usually taken to mean a very high level of spatial processing, involving a kind of survey representation of the environment, which makes it possible to move efficiently between the places charted on the map. Many disciplines have taken over this suggestive expression: psychology, of course, both animal and human, as well as ethology, neuroscience, architecture, town planning, geography, and artificial intelligence. In the literature on behavioural geography and cognitive psychology, cognitive mapping refers to the process of forming internal spatial representations of the environment that can be subsequently and flexibly used in navigating and in communicating with others, as opposed to the more rigid representations that make up pure route knowledge. The consensus view is that a cognitive map is more abstract than route knowledge and has image-like properties, thus allowing the agent to plan effective short-cuts and detours and to quickly estimate distances and bearings from any location within the map to any other such location (e.g., Downs and Stea, 1973; Golledge, 1987; O'Keefe and Nadel, 1978; Siegel and White, 1975).

Information used during the cognitive mapping process

Humans perform cognitive mapping both while travelling through environments and through the use of 'spatial products' (e.g., Liben, 1981), such as maps, photographs, videotapes, verbal (oral or written) descriptions, and, most recently, virtual environments. The literature provides abundant evidence that the nature of the cognitive map formed depends to some degree on the information available during knowledge acquisition (e.g., Thorndyke and Hayes-Roth, 1982). The focus of this section is on cognitive mapping based on travel through the environment.

Human travel occurs under a variety of circumstances. First, a person can be transported or otherwise passively guided through an environment. Second, the person can explore the environment without having any particular goal in mind. Third, the person might navigate through the environment seeking a particular destination. Navigation to a location outside the perceptual field requires that the person either already has some sort of knowledge about the environment or does so on the basis of some sort of spatial product (map, verbal instructions, etc.). Information acquired about the environment obviously depends on the circumstances of travel.

Information about the environment sensed directly or derived from the person's travel through it comes from the various sense modalities as well as from command signals issued to the muscle control system ('efference copy'). Vision provides the most information by far, both about the spatial layout of the near and distant environment and about spatial information contained in maps, aerial photographs, etc. Accordingly, most research on human cognitive mapping has focused on vision, but clearly some blind persons are able to navigate and to form cognitive maps of the environment using the remaining information available to them (audition, olfaction, touch, proprioception, and efference copy) (see Chapter 13).

A person travelling through an unfamiliar environment can form spatial representations of the environment using three distinct classes of information. The first of these is route information (Golledge, 1987; O'Keefe and Nadel, 1978; Siegel and White, 1975). Even if passively guided, the person can form associations between successive perceptual images along the route (for the visual sense, these images are generally referred to as 'views'). Alternatively, the person can store the route as a succession of segment lengths and turns. This type of 'route representation' is sufficient for repeating travel along the same route on subsequent occasions but by itself does not permit the computation of never experienced routes, such as short-cuts and detours. However, many such overlapping route representations densely covering an environment would provide the person with the capability of travelling from almost any location to almost any other.

The second type of information used in building spatial representations stems from the process known as 'path integration' (e.g., Etienne, 1992; Gallistel, 1990; Mittelstaedt and Mittelstaedt, 1982). Expressed in the most general terms, path integration is the process of updating one's current location on the basis of sensed displacements and turns (Loomis *et al.*, 1999) and is likely to involve an azimuthal reference (like the sun or a mountain range), vestibular inputs, and proprioception, possibly aided by optic flow and local features of the environment (Chance *et al.*, 1998; Klatzky *et al.*, 1998). Unlike route knowledge, path integration involves minimal storage of information in memory. For example, one minimal representation is the currently estimated location and heading within an allocentric frame (Mittelstaedt and Mittelstaedt, 1982; Müller and Wehner, 1988; Maurer and Séguinot, 1995). Used in connection with perceptual images of the environment, the position and orientation information obtained from path integration can be used to gradually develop a representation of the spatial disposition of different parts of the environment, a representation functionally equivalent to the common conception of cognitive map (Gallistel, 1990; Poucet, 1993; Thinus-Blanc, 1996).

The third type of information used in building spatial representations is based on landmark information (e.g., Benhamou, 1997, 1998; Poucet, 1993). A person exploring or navigating through novel territory can use bearing and distance information obtained from more distant landmarks, visible from different regions of the environment, as a means of integrating distinctive locations within the environment into a coherent global representation that, again, is functionally equivalent to the common notion of cognitive map.

In general, the information obtained from spatial products is vastly different from that obtained by travel through the environment. Viewing a map or an aerial photograph clearly provides direct information about the spatial disposition of places within the environment. Likewise, verbal information in the form of travel directions, etc., has little in common with the information obtained during actual travel. Despite these enormous differences in the type of information used for acquired environmental knowledge, the possibility exists that the resulting representations themselves are similar. This is the topic of the next section.

Functional properties of cognitive maps

Many studies have documented the elaboration of cognitive maps in humans (see for instance Golledge, 1987, for a review). According to Shemyakin (1962), people first construct 'route maps' and then acquire higher levels of spatial knowledge in the form of more detailed 'survey maps' which match the topography of the layout. Siegel and White (1975) and Pick and

Lockman (1981) make the distinction between three levels of spatial knowledge: (a) memorization of main landmarks (b) integration of those landmarks into a path or a sensorimotor sequence, and (c) elaboration of a survey-type representation where landmarks and paths are interconnected. Another distinction is given in terms of declarative or landmark knowledge, procedural or route knowledge, and configurational or survey (or abstract) knowledge. Reproducing a familiar route requires no more than a route-type mental representation, while selecting novel routes, like shortcuts, requires a more flexible representation, like a survey representation (Loomis *et al.*, 1993; Presson and Montello, 1994; Thinus-Blanc and Gaunet, 1997). Golledge (1987) suggests that the term 'map' should be interpreted more as a metaphor than a strict analogy; 'representations with map-like properties' is perhaps more appropriate. Indeed, cognitive maps are characterized by deformations and distortions (holes, folds, tears, cracks; see for instance Golledge, 1987; Tversky, 1981).

Recently, we (Thinus-Blanc, 1996; Thinus-Blanc and Gaunet, 1999) have proposed a dynamic sketch of spatial processing mechanism that assumes, in line with Neisser's hypothesis (1976) about a 'perceptual cycle', that the function of spatial representations is two-fold. First, of course, they are necessary for planning trajectories such as shortcuts and detours. But another important function is that these spatial representations control the organization of the spontaneous acquisition of information when an organism is confronted with a new situation. The variety of spatial representations acquired through development would, in adulthood, form 'schemas' possessing general properties common to all specific representations. The nature of such schemas would organize the acquisition of information when an organism is confronted with a new spatial situation. This hypothesis is based on data showing that behavioural regularities or exploratory patterns in a new environment differ between early blind persons and those who have had early visual experience (blindfolded sighted and adventitiously blind subjects). Accordingly, the strategy implemented by visually experienced participants is correlated with the best performance levels on further tests; whereas, the opposite is true for the early blind (Gaunet *et al.*, 1997; Gaunet and Thinus-Blanc, 1996). Thus, existing cognitive maps would shape the nature of newly constructed maps, which, in return, would modify and update already established maps. Indeed, this conception also applies to route knowledge, but given that this type of representation is simpler and less problematic, it is reasonable to think that the dynamic feature of this view is of less importance at this level. In addition, this conception puts emphasis on what is called 'local views', conceived as the building blocks of spatial processing. One of the consequences of locomotor activity, eye scanning, and head movements, all of which enlarge the zone of inspection, is the multiplication of different visual images. These images have been called 'local views' since they depend on a given position of the eyes, head, and body at any given moment. Together with movement-generated feedback cues (for example, path-integration), the perception of various local views of the same spatial arrangement leads to the extraction of spatial invariants that are independent of any particular direction of approach and would be charted on maps as abstract entities corresponding to places.

The relevance of VET for studying spatial cognition

Advantages and limits of VET

In humans, cognitive mapping research is usually based on static (and not on-line) measures of spatial behaviour in real or artificial settings: analyses of drawings, interviews, verbal reports, direction and distance estimates, and so on (for a review, see for instance Evans, 1980; Golledge, 1987). In natural situations, spatial behaviour and cognition can be modified by varying the exploration conditions, navigation tools, or exposure time. One major problem is the difficulty of controlling all possible parameters, such as spatial foreknowledge of the participants or the environmental features which are manipulated. There may also be problems in replicating experiments or making between-experiment comparisons because of differences in the experimental setting. It is mainly for these reasons that certain topics in human spatial cognition, such as navigation and wayfinding, have been studied using artificial representations of real and imaginary environments. However, laboratory studies are often unrealistic, and simulations (small-scale models, slide or film projections, etc.) have usually been static or dynamic but not (or only poorly) interactive (see for example Goldin and Thorndyke, 1982; O'Neill, 1992).

VET permits interactive navigation in a purely visual mode (Ellis, 1991; Durlach and Mavor, 1995; Kalawsky, 1993). Two main categories of systems are currently used: desk-top systems (which display the virtual environment on a fixed computer screen) and immersive-display systems (e.g., the environment is displayed on two small screens of a head-mounted display). Both have specific advantages and limitations, and their use is tightly related to the nature of questions under study. So far, there has been little research in the field using VET, but more and more people are becoming convinced of its relevance as a tool for spatial cognition studies (for reviews, see Darken et al., 1998; Loomis et al., in press; Péruch and Gaunet, 1998; Wilson, 1997). Some important virtues of VET are that they permit the creation of environments of varying complexity, provide continuous measurements during navigation, and afford the control of all of the spatial learning parameters: amount of exposure to the environment and the number, position, and nature of landmarks. Building a virtual maze and manipulating the above mentioned parameters can be relatively easy to do. Thus, the use of VET for studying spatial cognition offers many additional tools for experimentation and makes it possible to create experimental conditions that would be impossible in actual environments. As a result, the use of VET to investigate spatial processing pushes away the constraining limits of real world experimental situations.

However, it should also be mentioned that with the current state of the art, VET has many drawbacks: lack of realistic environmental modelling and image rendering, slow image generation, narrow field of view, poor spatial resolution, optical distortions, etc. Misperception of distance is frequently encountered in virtual environments (see for instance Loomis and Knapp, in press; Neale, 1996; Distler *et al.*, 1998; Witmer and Kline, 1998) and is probably the result of one or more of these drawbacks. Moreover, although devices with 3-D sound and haptic feedback are available, virtual environments are generally designed around just the visual modality. However, because researchers are progressively more aware of their requirements for research and because technical improvements are continually being made, suitable systems are becoming more and more common.

Although VET is relatively recent, cognitive mapping studies using it have already investigated several different issues. In brief, as is often the case when a new technology comes into existence, investigators have first tried to replicate (or/and extend) some 'basic' experiments that are concerned with cognitive mapping. An example is the study by Ruddle *et al.* (1997), in which most of the findings obtained by Thorndyke and Hayes-Roth (1982) in real settings were confirmed within virtual environments: participants were found to be equally good at performing direction and relative distance estimates. Such research confirms that the same processes operate in virtual (in general, purely visual) conditions and, at the same time, provides further validation of using VET for research in spatial cognition.

Cognitive mapping and VET

Here we give a brief overview of some issues in cognitive mapping that have been studied using VET (see Péruch and Gaunet, 1998, and Wilson, 1997, for more detailed accounts). One issue concerns the role of activity in linking visual percepts experienced while moving. Due to the fact that VET permits an interactive coupling of the participant's action and perception, some experiments have investigated the importance of the degree of interactivity between participants and a virtual environment while they are moving around in it.

Dynamic simulations can be experienced either passively (i.e., the user cannot choose the path through a simulation) or interactively (i.e., the user is allowed to choose the path). Some studies have shown the superiority of active over passive exploration (see for instance Christou and Bülthoff, 1998; Péruch *et al.*, 1995; Tong *et al.*, 1995), though discrepant results have also been found (Arthur, 1996; Satalich, 1995; Wilson *et al.*, 1997). At the present time, it is not obvious why spatial knowledge gained through spontaneous activity can be more precise than that acquired through passive exploration. The fact remains, however, that making decisions while getting acquainted with a new environment during active exploration implies the participant's involvement in the acquisition of spatial knowledge. He/she has to discriminate concurrently between those parts of the environment that have been already travelled from those that have not been. In this respect, it may be worth investigating to what extent poorer performance, that is often recorded after passive exploration, depends on the temporal organization of the travel. Designing a passive exploratory visit on the basis of spontaneous exploratory strategies of the best performers may help to better understand this problem. The use of VET appears to be especially relevant for such studies.

Another issue is related to the respective roles of information coming from the environment and of that related to locomotor activity (movement-generated feedback cues). Since vision is an important sense for spatial processing, it is possible first to concentrate on visual simulation and then to compare performance in a situation involving another sensory modality (audition, proprioception). In particular, such interactive simulations allow researchers to separately investigate the functional properties of two distinct processes usually involved in human navigation: piloting (which relies on position-fixing based on environmental cues) and path-integration (which involves updating one's estimate of current position based on integration of self-motion). In this respect, Péruch *et al.* (1997) have found, in a triangle completion task performed in purely visual conditions, poorer performance levels than in Loomis *et al.*'s (1993) study, in which the task was performed using walking without vision (with blindfolded participants). Other studies comparing spatial updating in real or virtual locomotion (see for example Chance *et al.*, 1998; Klatzky *et al.*, 1998; Gaunet *et al.*, submitted) have investigated the respective contribution of visual, vestibular, and proprioceptive information.

A further issue that is crucial for getting more insight into the cognitive map notion is related to the landmarks that are used to build spatial representations. For instance, Tlauka and Wilson (1994) have investigated to what extent landmarks are decisive in the acquisition of route knowledge in a virtual environment: performance was higher in the landmark group than in the non-landmark group. Gillner and Mallot (1998) have addressed the question of the effects of conditions of acquisition of spatial knowledge in a virtual maze with only local information (no global landmarks or compass information): subjects were able to learn the spatial configuration uniquely on the basis of sequences of local views and movements. In several experiments conducted in a virtual maze or 'arena' (Jacobs *et al.*, 1997, 1998), participants learn from different starting locations to find an invisible target that remains in a fixed location relative to distal cues. Removing distant landmarks had no important effect, while merging the same landmarks (changing their topographical relations) dramatically decreases performance. In a study conducted in a virtual arena with a hidden goal (as in Jacobs *et al.*, 1997, see above), with both landmarks and room geometry information available, Sandstrom *et al.* (1998) have found: (1) when the local landmarks within the arena used initially to learn the target location were removed or relocated while keeping the room geometry constant, performance worsened considerably; and (2) the disruptive effects of modifying the local landmark cues were much greater for females than for males. These experiments confirm a conceptual difference between 'proximal' and 'distal' spatial orientation, already made by Morris (1981) and others in the animal literature.

Some other studies have investigated the properties of the cognitive maps that are elaborated in virtual environments, through navigation and wayfinding tasks (for a discussion see also Darken *et al.*, 1998). With natural environments, navigation results in an orientation-free (or orientation-preferred) representation, while maps are orientation-dependent. Similar (see for instance Tlauka and Wilson, 1996) or contradictory (see for instance May *et al.*, 1995; Péruch and Lapin, 1993; Rossano and Moak, 1998; Richardson *et al.*, 1999) results have been found with navigation in virtual environments.

The transfer of spatial knowledge from virtual to real environments is another important issue investigated in some studies. They demonstrate that virtual environments contain much of the essential spatial information that is utilized by people in real environments. Regian *et al.* (1992) showed that subjects who had navigated in a virtual maze performed a wayfinding task in a real environment better than expected on the basis of chance, revealing a transfer of spatial information. Witmer *et al.* (1996) found a better transfer of route knowledge to real environments from virtual ones than from verbal directions and photographs. Bliss *et al.* (1997) found that firefighters trained to navigate a rescue route in an unfamiliar building performed equally well after virtual navigation and after map examination, with both being better than those who had not been trained. Wilson *et al.* (1997b) showed evidence of a transfer of learning from a virtual to a real environment for pointing and map drawing tasks, while transfer for (Euclidean and route) distance estimates was less evident. A study by Waller *et al.* (1998) showed that extensive training in a virtual environment eventually surpassed real world training. Finally, a recent work by Darken and Banker (1998) showed that intermediate users benefit more than either advanced or beginner users from the virtual environment training; moreover, the level of ability seems to be more important than the training method to perform navigation.

In summary, important theoretical issues related to the nature and properties of cognitive maps can be tackled using VET. Obviously, this technology, like all preceding ones, will not solve once and for all the remaining crucial problems related to spatial cognition, but, by opening new methodological perspectives, it allows researchers to sharpen the questions they ask through experimental designs.

Combining VET and brain imagery: a recent approach to cognitive mapping

A consistent trend in the evolution of cognitive science and neuroscience is the development of fruitful interactions between the two fields. On the one hand, studying the brain structures involved in cognitive processes in general and spatial representations in specific situations provides useful tools for further analysing and understanding the psychological mechanisms involved in such processing. On the other hand, data deriving from the study of psychological processes in relation to brain functioning provide invaluable information to those involved in clinical approaches and re-education of patients with brain lesions. A major example of the interest in such interactions is that provided by O'Keefe and Nadel in their famous book *The Hippocampus as a Cognitive Map* (1978). They proposed a cognitive mapping theory of hippocampal functioning that is still raising a huge interest among neuro-biologists and psychologists, and has been confirmed by a variety of studies in the field of cognitive neuroscience. However, other brain regions, such as the posterior parietal and the prefrontal cortices are also primarily involved in spatial processing. The main problems at issue consist in defining their specific respective roles and in delineating how they interact during the various phases of spatial processing.

So far, most of the data have come from animal studies and from human clinical studies, such as 'case reports'. However, the recent development and accessibility of techniques of brain imaging allow researchers to conduct more systematic studies of human brain functioning on homogeneous groups of intact subjects. Very recently, the combined use of brain imagery and VET appears to be of major interest for studying spatial cognition at both behavioural and brain levels. Each brain imagery method has specific complementary advantages. Event Related Potentials (ERPs) and Magnetic Electroencephalography (MEG) methods have very good temporal resolution (0–300 ms) but provide only rough indications about the limits of the brain regions that are active at a given moment. In contrast, temporal resolution is poor (1 min) with Positron Emission Tomography (PET) method but the spatial resolution is very high (1–2 mm). Finally, functional Magnetic Resonance Imagery (fMRI) seems to provide relatively high resolution for both temporal (1 sec) and spatial (1 mm) parameters.

Most of the studies on human spatial cognition have been conducted using PET, but fMRI becomes more and more attractive for the reasons mentioned above. For instance, in a PET study by Flitman *et al.* (1997), adults had to perform tasks in two control conditions (motor and visual) and in two experimental conditions – an uncovered maze and a partially covered maze (only locally visible). The results indicated that maze processing lateralizes to the right hemisphere, with different areas activated in the two maze conditions. These data are in agreement with conclusions of many other studies showing that, unlike patients with a left brain damage, those with a right brain lesion are deeply impaired in spatial processing.

The route/map issue has been investigated by Maguire *et al.* (1997) in a study involving the recall of complex routes around London by taxi drivers. Compared with baseline and other non-topographical memory tasks, this task of topographical memory retrieval activates a network of brain regions, including the right hippocampus. In contrast, the right hippocampus is not activated during the recall of famous landmarks for which subjects have no knowledge of their location within a spatial framework. One main conclusion drawn by the authors is that the right hippocampus is recruited specifically for navigation in large-scale spatial environments. Finally, Ghaem *et al.* (1997) go further in their analysis. They have shown that the mental simulation of a route which had been previously learned by actual navigation is subserved by two distinct networks: a non-specific memory network including the posterior and middle parts of the hippocampal regions, the dorsolateral prefrontal cortex and the posterior cingulum; and a specific mental navigation network, comprising the left precuneus, insula and medial part of the hippocampal regions.

Up until now, only a few brain imaging studies have been conducted in combination with VET. The role of landmarks in topographic memory was examined in virtual environments in two PET experiments by Maguire *et al.* (1998b). In the first experiment, the environment contained salient objects and textures that could be used to discriminate different rooms. In the second experiment, the environment was empty, rooms being distinguishable only by their shape. In both experiments, learning activated a network of bilateral occipital, medial parietal and occipito-temporal regions. However, the presence of salient objects and textures in the environment (first experiment) additionally resulted in increased activity in the right parahippocampal gyrus, while this region was not activated during exploration of the empty environment (second experiment). These data indicate the presence of an area that specifically processes landmarks, thus confirming psychological studies suggesting that landmarks have a special status in spatial cognition. Another PET experiment by Maguire *et al.* (1998a) compared three conditions of virtual navigation. In the first condition the subjects could head directly toward the goal; in the second condition barriers precluded direct routes, necessitating the use of detours, while in the third condition the subjects moved through the town following arrows (no internal representation was needed). Successful trials in both direct and detour wayfinding (first and second conditions) involved the right hippocampus, which was not the case in the third condition. Moreover, only successful navigation requiring detours (second condition) activated the left frontal gyrus. This study again stresses the role of the hippocampus in high level spatial representations. With fMRI, an experiment by Aguirre *et al.* (1996) examined the contributions of the hippocampus and of the parahippocampal areas to the learning and recall of topographic information in a virtual maze. Data showed that in the human (contrary to the rat but similarly to the monkey) this function seems to be not located in the hippocampus itself, but rather in the parahippocampal gyri and in some cortical areas known to project to the parahippocampus. In another study by Aguirre and D'Esposito (1997), participants made judgements regarding the appearance and position of familiar locations in a virtual environment (landmark versus survey information). A dorsal/ventral dissociation was found: a greater activity in the parahippocampus, in the fusiform/lingual gyrus, and in the middle right occipital gyrus was observed during the appearance task, while the position task activated the inferior parietal lobule, the precuneus, the superior parietal lobule, and premotor cortex on the left. Such a dissociation confirms that environmental knowledge is not represented by a unitary system: instead, it is functionally distributed across the neocortex.

Future research on cognitive mapping with new technologies

In the previous section, we have provided some examples that demonstrate the relevance of some new technologies such as VET and functional brain imagery for spatial cognition and cognitive mapping research. It must be emphasized that these technologies are quite recent and that technical progress will be made that will open even newer possibilities. With respect to VET, more and more complex and realistic models of environments will be available. They should allow researchers to better define optimal conditions of transfer of spatial knowledge from virtual to real environments, and, in so doing, reveal which spatial information is important for cognitive mapping and why. Moreover, the combination of VET with other techniques related to movement and/or navigation is promising. For instance, the development of whole body interfaces (or at least more natural movement interfaces) such as motion platforms, treadmills, and stationary bicycles will allow a more realistic simulation of locomotion and navigation, and in combination with VET should open a range of possibilities. In Tuebingen, movement in virtual environments is produced by a 'virtual bicycle' (see for instance Distler *et al.*, 1998). Another example is the combination of VET with the Global Positioning System, allowing subjects to walk naturally through large-scale virtual environments (see Loomis *et al.*, 1998 for an example in connection with auditory virtual environments). In this respect an important issue which is beyond the scope of this chapter is the domain of applications. There are strong potential benefits of VET for instance for the rehabilitation of disabled individuals (blind, aged, brain damaged, etc.) and for human–computer interaction devices.

Important issues in spatial cognition can be studied efficiently using VET since, as underlined in the previous section, this technology permits experimental designs that would be impossible to implement in the real world. For instance, the status and role of landmarks in the building up of spatial representations is an important question that has been relatively neglected so far. What is a landmark? How is it used? Why is a cue, among so many others, selected and used as a landmark for cognitive mapping and navigation? Indeed, this issue should benefit from the possibility of easily bringing a wide range of modifications to a virtual environment (nature and number of landmarks). Another important issue is that related to the respective contribution of visual and non-visual movement-generated information. Since both types of inputs are at the basis of cognitive mapping, it is important to better understand how they contribute to such a process, whether their effects are additive or interact, etc. The capability of dissociating them using VET should constitute a significant advancement. Moreover, the more specific question of the use of distant versus proximal landmarks in navigation appears also an important issue likely to progress with VET, since distances of the landmarks can be controlled and manipulated.

Concerning functional brain imaging, most of the data obtained so far rely on the use of PET, but this is becoming less and less true. However, it is obvious that the combination of several techniques (e.g., fMRI and ERPs) should allow us to go further than simply locating the structures involved in cognitive mapping by combining a fine-grained spatial and temporal analysis. The network of structures that have specific functions in the various levels and phases of spatial processing should be made apparent, along with the temporal organization of the recruitment of the various brain regions involved. Finally, the combination of VET and functional brain imaging techniques (see for instance Maguire *et al.*, 1998a) is very promising since it is a means to free the participant from some of the constraints inherent in the recording devices. For instance, PET and fMRI require the subject's head to remain perfectly still while recording. With VET, however, the subject has the possibility of moving just the fingers to manually control travel through a virtual environment. The visual consequences of these movements appear on a screen in front of the subject. There is an interactivity and a sense of movement, despite remaining totally stationary within the laboratory. Finally, the study of differences between males and females while performing spatial tasks is of interest not only at the purely psychological level, but also in relation to recent studies in rats showing a strong hormonal influence on brain functioning (e.g., McEwen *et al.*, 1997). This issue needs to be tackled in humans in order to elucidate the often observed differences in performance between females and males. Combining brain imagery and VET appears to be a promising tool in order to investigate differences in brain functioning while performing spatial tasks in both small- and large-scale virtual situations.

Acknowledgements

This chapter was written while Jack Loomis was visiting the Center for Research in Cognitive Neuroscience in Marseille on a grant from the Centre National de la Recherche Scientifique, France. Parts of this chapter were presented at the International Workshop 'Self-organization, cognitive mapping, urban and regional systems, and spatial information' held at Maison des Sciences de l'Homme, Paris, May 17–18, 1998.

References

Aguirre, G.K., Detre, J.A., Alsop, D.C. and D'Esposito, M. (1996) The parahippocampus subserves topographical learning in man. *Cerebral Cortex*, 6, 823–9.

—— and D'Esposito, M. (1997) Environmental knowledge is subserved by separable dorsal/ventral neural areas. *The Journal of Neuroscience*, 17, 2512–18.

Arthur, E.J. (1996) Orientation specificity in the mental representation of three dimensional environments. Unpublished doctoral dissertation, University of Minnesota.

Benhamou, S. (1997) On systems of reference involved in spatial memory. *Behavioural Processes*, 40, 149–63.

—— (1998) Place navigation in mammals: a configuration-based model. *Animal Cognition*, 1, 55–63.

Bliss, J.P., Tidwell, P.D. and Guest, M.A. (1997) The effectiveness of virtual reality for administering spatial navigation training for firefighters. *Presence*, 6, 73–86.

Chance, S.S., Gaunet, F., Beall, A. and Loomis, J.M. (1998) Locomotion mode affects the apprehension of spatial layout: The contribution of vestibular and proprioceptive inputs. *Presence: Special Issue on Spatial Navigation*, 7, 168–78.

Christou, C.G. and Bülthoff, H.H. (1998) Using virtual environments to study spatial encoding. *Proceedings of the Spatial Cognition Conference*, Dublin, 17–18 August.

Darken, R.P., Allard, T. and Achille, L.B. (1998) Spatial orientation and wayfinding in large-scale virtual spaces: An introduction. *Presence: Special Issue on Spatial Navigation*, 7, 101–7.

—— and Banker, W.P. (1998) Navigating in natural environments: a virtual environment training transfer study. *Proceedings of VRAIS '98*, 12–19.

Distler, H.K., Van Veen, H.A., Braun, S.J., Heinz, W., Franz, M.O. and Bülthoff, H.H. (1998) Navigation in real and virtual environments: Judging orientation and distance in a large-scale landscape. In Goebel, M., Lang, U., Landauer, J. and Walper, M. (eds), *Eurographics Workshop Proceedings Series*, #42.

Downs, R.M. and Stea, D. (eds), (1973) *Image and Environment. Cognitive Mapping and Spatial Behavior*. Chicago, IL: Aldine.

Durlach, N.I. and Mavor, A.S. (eds), (1995) *Virtual Reality: Scientific and Technological Challenges*. Washington, DC: National Academy.

Ellis, S.R. (1991) *Pictorial Communication in Virtual and Real Environments*. London: Taylor and Francis.

Etienne, A.S. (1992) Navigation of a small mammal by dead reckoning and local cues. *Current Directions in Psychological Science*, 1, 48–52.

Evans, G.W. (1980) Environmental cognition. *Psychological Bulletin*, 88, 259–87.

Flitman, S., O'Grady, J., Cooper, V. and Grafman, J. (1997) PET imaging of maze processing. *Neuropsychologia*, 35, 409–20.

Gallistel, C.R. (1990) *The Organization of Learning*. Cambridge, MA: MIT.

Gaunet, F., Martinez, J.L. and Thinus-Blanc, C. (1997) Early-blind subjects' spatial representation of manipulative space: exploratory strategies and reaction to change. *Perception*, 26, 345–66.

Gaunet, F., Péruch, P. and Thinus-Blanc, C. (submitted) Role of visual and non-visual movement-related information in the acquisition and use of spatial representations.

—— and Thinus-Blanc, C. (1996) Early-blind subjects' spatial abilities in the locomotor space: Exploratory strategies and reaction-to-change performance. *Perception*, 25, 967–81.

Ghaem, O., Mellet, E., Crivello, F., Tzourio, N., Mazoyer, B., Berthoz, A. and Denis, M. (1997) Mental navigation along memorized routes activates the hippocampus, precuneus, and insula. *NeuroReport*, 8, 739–44.

Gillner S. and Mallot, H.A. (1998) Navigation and acquisition of spatial knowledge in a virtual maze. *Journal of Cognitive Neuroscience*, 10, 445–63.

Goldin, S.E. and Thorndyke, P.W. (1982) Simulating navigation for spatial knowledge acquisition. *Human Factors*, 24, 457–71.

Golledge, R.G. (1987) Environmental cognition. In Stokols, D. and Altman, I. (eds), *Handbook of Environmental Psychology*, vol. 1. New York: Wiley, pp. 131–74.

Jacobs, W.J., Laurance, H.E. and Thomas, K.G.F. (1997) Place learning in virtual space I: Acquisition, overshadowing, and transfer. *Learning and Motivation*, 28, 521–41.

—— Thomas, K.G.F., Laurance, H.E. and Nadel, L. (1998) Place learning in virtual space II: Topographical relations as one dimension of stimulus control. *Learning and Motivation*, 29, 288–308.

Kalawsky, R.S. (1993) *The Science of Virtual Reality and Virtual Environments*. New York: Addison-Wesley.

Klatzky, R.L., Loomis, J.M., Beall, A.C., Chance, S.S. and Golledge, R.G. (1998) Spatial updating of self-position and orientation during real, imagined, and virtual locomotion. *Psychological Science*, 9, 293–8.

Liben, L.S. (1981) Spatial representation and behavior: Multiple perspectives. In Liben, L.S., Patterson, A.H. and Newcombe, N. (eds), *Spatial Representation and Behavior across the Life Span. Theory and Application*. New York: Academic, pp. 3–36.

Loomis, J.M., Blascovich, J.J. and Beall, A.C. (in press) Immersive virtual environment technology as a basic research tool in psychology. *Behavior Research Methods, Instruments, & Computers*.

—— Golledge, R.G. and Klatzky, R.L. (1998) Navigation system for the blind: Auditory display modes and guidance. *Presence: Special Issue on Spatial Navigation*, 7, 193–203.

—— Klatzky, R.L., Golledge, R.G., Cicinelli, J.G., Pellegrino, J.W. and Fry, P.A. (1993) Nonvisual navigation by blind and sighted: Assessment of path integration ability. *Journal of Experimental Psychology: General*, 122 (1), 73–91.

—— Klatzky, R.L., Golledge, R.G. and Philbeck, J.W. (1999) Human navigation by path integration. In Golledge, R.G. (ed.), *Wayfinding: Cognitive Mapping and Spatial Behavior*. Baltimore, MD: Johns Hopkins, pp. 126–51.

—— and Knapp, J.M. (in press) Visual perception of egocentric distance in real and virtual environments. In Hettinger, L.J. and Haas, M.W. (eds), *Virtual and Adaptive Environments*. Hillsdale, NJ: Erlbaum.

McEwen, B.S., Alves, S.E., Bulloch, K. and Weiland, N.G. (1997) Ovarian steroids and the brain: Implications for cognition and aging. *Neurologia*, 48, S8–15.

Maguire, E.A., Burgess, N., Donnett, J.G., Frackowiak, R.S.J., Frith, C.D. and O'Keefe, J. (1998a) Knowing where and getting there: A human navigation network. *Science*, 280, 921–4.

——— Frackowiak, R.S.J. and Frith, C.D. (1997) Recalling routes around London: Activation of the right hippocampus in taxi drivers. *The Journal of Neuroscience*, 17, 7103–10.

——— Frith, C.D., Burgess, N., Donnett, J.G. and O'Keefe, J. (1998b) Knowing where things are: Parahippocampal involvement in encoding object locations in virtual large-scale space. *Journal of Cognitive Neuroscience*, 10, 61–76.

Maurer, R., and Séguinot, V. (1995) What is modelling for? A critical review of the models of path integration. *Journal of Theoretical Biology*, 175, 457–75.

May, M., Péruch, P. and Savoyant, A. (1995) Navigating in a virtual environment with map-acquired knowledge: Encoding and alignment effects. *Ecological Psychology*, 7, 21–36.

Mittelstaedt, H. and Mittelstaedt, M.L. (1982) Homing by path integration. In Papi, F. and Wallraff, H.G. (eds), *Avian Navigation*. New York: Springer, pp. 126–51.

Morris, R.G.M. (1981) Spatial localization does not require the presence of local cues. *Learning and Motivation*, 12, 239–60.

Müller, M., and Wehner, R. (1988) Path integration in desert ants, Cataglyphis fortis. *Proceedings of the National Academy of Sciences*, 85, 5287–90.

Neale, D.C. (1996) Spatial perception in desktop virtual environments. In *Proceedings of the 40th annual meeting of the Human Factors and Ergonomics Society*. Santa Monica, CA: Human Factors and Ergonomics Society.

Neisser, U. (1976) *Cognition and Reality*. San Francisco, CA: W. H. Freeman.

O'Keefe, J. and Nadel, L. (1978) *The Hippocampus as a Cognitive Map*. Oxford: Oxford University Press.

O'Neill, M. (1992) Effects of familiarity and plan complexity on wayfinding in simulated buildings. *Journal of Environmental Psychology*, 12, 319–27.

Péruch, P. and Gaunet, F. (1998) Virtual environments as a promising tool for investigating human spatial cognition. *Cahiers de Psychologie Cognitive/Current Psychology of Cognition*, 17, 881–99.

——— and Lapin, E. (1993) Route knowledge in different spatial frames of reference. *Acta Psychologica*, 84, 253–69.

——— May, M. and Wartenberg, F. (1997) Homing in virtual environments: Effects of field of view and path layout. *Perception*, 26, 301–11.

——— Vercher, J.L. and Gauthier, G.M. (1995) Acquisition of spatial knowledge through visual exploration of simulated environments. *Ecological Psychology*, 7, 1–20.

Pick, H.L., Jr. and Lockman, J.J. (1981) From frames of reference to spatial representations. In Liben, L.S., Patterson, A.H. and Newcombe, N. (eds), *Spatial Representation and Behavior across the Life Span*. New York: Academic, pp. 39–61.

Poucet, B. (1993) Spatial cognitive maps in animals: New hypotheses on their structure and neural mechanisms. *Psychological Review*, 100, 163–82.

Presson, C.C. and Montello, D.R. (1994) Updating after rotational and translational body movements: Coordinate structure of perspective space. *Perception*, 23, 1447–55.

Regian, J.W., Shebilske, W.L. and Monk, J.M. (1992) Virtual reality: An instructional medium for visual-spatial tasks. *Journal of Communication*, 42, 136–49.

Richardson, A.E., Montello, D.R. and Hegarty, M. (1999) Spatial knowledge acquisition from maps, and from navigation in real and virtual environments. *Memory and Cognition*, 27, 741–50.

Rossano, M.J. and Moak, J. (1998) Spatial representations acquired from computer models: Cognitive load, orientation specificity and the acquisition of survey knowledge. *British Journal of Psychology*, 89, 481–97.

Ruddle, R.A., Payne, S.J. and Jones, D.M. (1997) Navigating buildings in 'desktop' virtual environments: Experimental investigations using extended navigational experience. *Journal of Experimental Psychology: Applied*, 3, 143–59.

Sandstrom, N.J., Kaufman, J. and Huettel, S.A. (1998) Males and females use different distal cues in a virtual environment navigation task. *Cognitive Brain Research*, 6, 351–60.

Satalich, G.A. (1995) Navigation and wayfinding in virtual reality: Finding proper tools and cues to enhance navigation awareness. Unpublished MA thesis, University of Washington.

Shemyakin, F.N. (1962) Orientation in space. In Ananyev, B.G. *et al.* (eds), *Psychological Science in the USSR*, vol. 1, Washington, DC: Office of Technical Services, pp. 186–255.

Siegel, A.W. and White, S.H. (1975) The development of spatial representations of large-scale environments. In Reese, H.W. (ed.), *Advances in Child Development and Behavior*, vol. 10. New York: Academic, pp. 9–55.

Thinus-Blanc, C. (1996) *Animal Spatial Cognition: Behavioral and Brain Approaches*. Singapore: World Scientific.

—— and Gaunet, F. (1997) Representations of space in blind persons: Vision as a spatial sense? *Psychological Bulletin*, 121, 20–42.

—— and —— (1999) The organizing function of spatial representations. Spatial processing in animal and man. In Golledge, R.G. (ed.), *Wayfinding: Cognitive Mapping and Spatial Behavior*. Baltimore, MD: Johns Hopkins, pp. 294–307.

Thorndyke, P.W. and Hayes-Roth, B. (1982) Differences in spatial knowledge acquired from maps and navigation. *Cognitive Psychology*, 14, 560–89.

Tlauka, M. and Wilson, P.N. (1994) The effects of landmarks on route-learning in a computer-simulated environment. *Journal of Environmental Psychology*, 14, 305–13.

—— and —— (1996) Orientation-free representations from navigation through a computer simulated environment. *Environment and Behavior*, 28, 647–64.

Tolman, E.C. (1948) Cognitive maps in rats and men. *Psychological Review*, 55, 189–208.

Tong, F.H., Marlin, S.G. and Frost, B.J. (1995) Visual-motor integration and spatial representation in a visual virtual environment. *Investigative Ophthalmology and Visual Science*, 36, 1679.

Tversky, B. (1981) Distortions in memory for maps. *Cognitive Psychology*, 13, 407–33.

Waller, D., Hunt, E., and Knapp, D. (1998) The transfer of spatial knowledge in virtual environment training. *Presence: Special Issue on Spatial Navigation*, 7, 129–43.

Wilson, P.N. (1997) Use of virtual reality computing in spatial learning research. In Foreman, N. and Gillett, R. (eds), *Handbook of Spatial Research Paradigms and Methodologies. Vol. 1: Spatial Cognition in the Child and Adult*. Brighton: Psychology Press, pp. 181–206.

—— Foreman, N., Gillett, R. and Stanton, D. (1997a) Active versus passive processing of spatial information in a computer simulated environment. *Ecological Psychology*, 9, 207–22.

—— , —— and Tlauka, M. (1997b) Transfer of spatial information from a virtual to a real environment. *Human Factors*, 39, 526–31.

Witmer, B.G., Bailey, J.H. and Knerr, B.W. (1996) Virtual spaces and real world places: Transfer of route knowledge. *International Journal of Human-Computer Studies*, 45, 412–28.

—— and Kline, P.B. (1998) Judging perceived and traversed distance in virtual environments. *Presence: Special Issue on Spatial Navigation*, 7, 144–67.

8   Micro- and macro-scale environments

Scott Freundschuh

Introduction

Nothing distinguishes more the spatial cognitive research of psychologists from that of geographers than the size of the environment that each discipline conducts its empirical work (Blaut, 1999). Though the research of each discipline does cross the macro–micro boundary from time to time, prevailing practice indicates that spatio-cognitive empirical research in psychology is conducted in micro-scale spaces, whereas similar research in geography is conducted in macro-scale spaces. Canter (1977) captured the nature of this 'dichotomist interest' when he contrasted the experimental laboratory space of psychologists, which he called 'object space', to the outside laboratory space of geographers, which he called the 'space of place'. 'While geographers operate at a regional scale, psychologists operate at the scale of a building' (Freundschuh and Egenhofer, 1997, p. 364). At the recent Specialist's Meeting of the National Center for Geographic Information and Analysis, 'Scale and Detail in the Cognition of Geographic Information', this fundamental disciplinary difference in perspective was affirmed. Discussions at this meeting often led participants to the 'micro–macro wall', with psychologists and geographers on somewhat opposite sides when discussing the sorts of environments we 'should' be using in empirical studies. The view from psychology tends to give little credence to the larger, complex environments that geographers are more interested in testing, pointing out that there are too many confounding variables that cannot be controlled in these types of experimental spaces. On the other hand, the view from geography questions the applicability of studies carried out in small, 'simple' environments (such as a table-top space) to the cognition of larger geographic spaces (Freundschuh and Egenhofer, 1997; Montello, 1993).

Figure 8.1 is a cartoon that was penned during discussions at this specialist meeting about the sorts of maps that psychologists and geographers use for empirical research (Montello and Golledge, 1999, Figure F). This cartoon, illustrating the difference between maps that psychologists and geographers use in experiments, is indicative of the contrast between

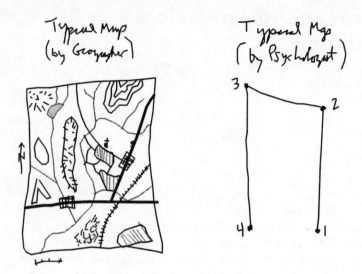

Figure 8.1 Typical map by geographers and psychologists.

Source: Adapted from Montello and Golledge, 1999, Figure F, p. 65.

geographers' and psychologists' apparent 'view of the world' when conducting empirical studies.

The map on the left illustrates an imaginary space, that of a geographic region, representing real world features such as roads, bridges, rivers and lakes, fields, railroad tracks, a cemetery, and some indication of relief. Geographers are interested in observing the results of spatio-cognitive processes and behaviour in a geographic context – that of a large-scale or geographic space rich in complexity, information and detail, and more typical of everyday experience. The map on the right illustrates a simple relationship between four nodes and three lines in space. One could posit that the four nodes represent cities and that the three lines represent roads, but the representation is not reflective of what is generally considered a map by both the general public and geographers (see Vasiliev *et al.*, 1990 for a discussion of 'what is a map?'). At most, it is a simple map. Though simple, this sort of testing environment affords psychologists the possibility of observing spatio-cognitive processes and behaviours in a controlled experimental environment that is less likely to be influenced by confounding influences. Both perspectives are useful and necessary in cognitive mapping research.

Still divided

In spite of considerable interaction between geographers and psychologists over the past thirty years with regard to cognitive mapping research (e.g.,

Downs and Stea, 1973; Moore and Golledge, 1976; Gärling and Golledge, 1993; Montello and Freundschuh, 1995), it appears that little has been done to bridge this divide in experimental spaces. Recent literature indicates growing interest in understanding the effect that scale and/or size of a space has on spatial cognition (Blaut, 1999; Frank, 1996; Freundschuh and Egenhofer, 1997; Montello and Golledge, 1999; Tversky *et al.*, 1999). The impetus for this recent body of literature is to understand how people conceptualize and understand space, thereby (either directly or indirectly) developing a general model for spatial representation (Abler, 1987). The knowledge gained through the development of a general model of spatial representation is, in the context of current research in geographic information science, intended to foster the design and development of geographic information systems (GIS). The goal of this development in GIS, in the broadest sense of the technology, is to create GIS that support human spatial cognition by providing systems that are 'user friendly' in all aspects of spatial data analysis (NCGIA, 1988).

Spatial cognition occurs in a variety of spaces, for example, table-top, environmental, geographical, and map space (Freundschuh and Egenhofer, 1997). Each type provides a different *context* in which spatial relationships can be described and understood (for a review of space types, see Freundschuh and Egenhofer, 1997; and Tversky *et al.*, 1999). The focus of this chapter is to provide a broader perspective on types of spaces outside of the micro/macro distinction, exploring the literature for evidence of a more diverse view of spaces. This chapter, like others in this volume, reviews past and present research that have been key in this area, and provides insights about the directions research in this area of cognitive mapping might go.

Scale versus size of space

Researchers have imbued the word scale with different meanings, leading to an on-going confusion of the notions of scale and size. Scale has been used to mean absolute size, relative size, resolution, granularity, and detail (Montello and Golledge, 1999). A prime example of this confusion is the use of the terms large- and small-scale. In psychology, small-scale means a space small in absolute size, whereas a large-scale space is one that is large in absolute size. In contrast, small-scale in geography means the representation of a large space on a map of relatively small absolute size, whereas large-scale means the representation of a small space on a map. In this chapter, the term size will refer to the absolute size of a space; for example, the dimension of a room in square metres, the size of a city measured in square kilometres, or the physical size of a map measured in square centimetres. In contrast, scale will be used to explain the relation between two spaces of different absolute sizes. Therefore, the scale of a map (or to some other map-like representation such as a model or an aerial photograph) refers to the representative fraction that describes the relation between distances on the map and the corresponding real world distance.

Evolution of models of space

Early research that distinguished between different kinds of spaces included two primitive types: small and large spaces (Canter, 1977; Downs and Stea, 1977; Ittelson, 1973; Kuipers, 1978; Lynch, 1960). In this research, small spaces (referred to as small-scale) are spaces where one can see all places within that space from one vantage point. Typical examples of these kinds of spaces include room size spaces and smaller. Large spaces (called large-scale), on the other hand, are spaces which cannot be perceived from one single view, and therefore, require locomotion to experience them. Large spaces, by necessity, are learned piecemeal over time, and include spaces ranging from inside building to city size spaces.

After this initial work in the 1960s and 1970s, models evolved to include three kinds of space that were essentially a finer-grain dissection of the previous models. Gärling and Golledge (1987) and Mandler (1983) included in their models *small*, *medium* and *large* size spaces, ranging from 'table-top' in size, to spaces the size of countries. Siegel (1981) also included three space types in this model, ranging from small to large spaces, as well as very large spaces that he asserted are represented on *maps* because first hand experience is not practical or feasible. Siegel was the first to explicitly suggest that maps are a unique space in and of themselves, and should be considered differently from other kinds of spaces. Both Kuipers (1978) and Mandler (1983) also implied a unique role for maps previously, but neither of their models specifically included maps as a space type.

Zubin (1989) expanded the thinking on kinds spaces by including in his model *A-spaces* and *C-spaces*. A-spaces are small, manipulable object spaces that are smaller than the body, are viewed from one perspective, and can be held, turned, rotated, etc. C-spaces are essentially an expansion on what was meant by small space in previous research, to include not only room size spaces that can be seen from one vantage point, but to include for example a large auditorium, a scenic overlook, the horizon, and a view from a plane. These spaces are perceived by panning the landscape.

Montello (1993) provided a distinction in large-size spaces that resulted in what he called *environmental spaces* and *geographical spaces*. Environmental spaces surround and engulf, and can only be experienced via locomotion through the space. These spaces are learned through repeated experience over time, and include buildings, neighbourhoods, and cities. In contrast, geographical spaces are much larger than environmental spaces, so much so that they cannot be perceived from direct experience. These spaces, which include states, countries and the solar system, are experienced via symbolic representations such as maps and 3-D models.

Current model of structures of space

From this theoretical framework of types of spaces provided in the literature, and considering the properties of manipulability, locomotion, and size of space, Freundschuh and Egenhofer (1997) proposed a classification of space that included six different types of space. The property of *manipulability* refers to the ability to grasp, turn, and move objects in space. Spaces are comprised of objects (or features) that are either manipulable, or non-manipulable. In this sense, space is established by a configuration of objects that people interact with, or among. The property of *locomotion* refers to the necessity to travel through a space in order to perceive or experience, and subsequently learn the space. These spaces are viewed from many perspectives, and therefore are orientation free. Spaces either require locomotion to experience them, or not. The property of *size* considers how spatial experience is constrained by size of space. The nature of the relationships among these three properties is used as a basis for defining different types of geographic spaces. Two of the properties have binary, mutually exclusive values (manipulability and locomotion), while one property has binary, possibly inclusive values (size). A systematic and complete combination of these properties gives rise to different combinations of properties that may exist between manipulability, locomotion, and size. The six types of spaces defined by Freundschuh and Egenhofer (1997) include:

- *manipulable object space*: comprises objects smaller than the human body, that can be held, turned, rotated, etc., and do not require locomotion to experience them;
- *non-manipulable object space*: comprises objects that are non-manipulable, typically larger than the human body and smaller than a house, and require locomotion to view all parts of the object;
- *environmental space*: spaces that require locomotion to experience them, and are learned piecemeal over time, such as inside of building spaces, to neighbourhoods and city size spaces;
- *geographic space*: very large spaces, such as cities, states, countries and the universe, that due to practical limitations, cannot be perceived in their entirety via locomotion;
- *panoramic space*: small to large spaces, including views in a room, an auditorium, a large field, and form a scenic overlook, that can be viewed from one vantage point by scanning or panning the space;
- *map space*: includes the representation of potentially all spaces, but usually large spaces, creating instances of large size space in a small size space. These are symbolic representations that are the result of cartographic generalization aimed at simplifying spatial information.

These various types of space that have been explored in the cognitive mapping literature offer a basis for re-thinking spatial cognition and spatial behaviour. They offer new insights into the kinds of spaces people live in, and provide a theoretical basis for how different spaces might shape and mould spatial cognitive representations. What is lacking in the literature, though, is broad-based evidence that these six different spaces are truly unique, empirically distinguishable, and do in fact result in different cognitive representations. There are a handful of studies that offer some mechanisms and evidence that enable a differentiation between spaces that result in different cognitive representations. Some of this literature attempts to formalize, mathematically, methods for differentiating between different kinds of space, whereas other work is empirical in nature, attempting to find clinical evidence for the existence of different spaces.

Formalisms of perceptual and cognitive space

Couclelis and Gale (1986) argued a formalization of spaces based on the algebraic structure of Abelian groups. The authors considered five axioms for identifying different types of space: the closure law [for all a, b that belong to S, $a*b$ belong to S]; the associative law [for all a, b, c that belong to S $(a*b)*c = a*(b*c)$]; the identity element [for all a that belong to S, there exists an element e that belongs to S such that $a*e - e*a = a$]; the inverse law [for every a that belongs to S, there exists an element b that belongs to S such that $a*b = b*a = e$]; and the commutative law [for all a, b that belong to S, $a*b = b*a$] – for identifying different types of spaces. The main goal of their thesis was to distinguish between *perceptual space* (small size spaces) and *cognitive space* (large size spaces). Couclelis and Gale (1986) defined perceptual space as what can be seen or observed through the senses at one time, and defined cognitive space as spaces that cannot be captured immediately with our sensors, and therefore must be integrated over time to be understood. Though their main goal was to distinguish between perceptual and cognitive space, the authors distinguished between a total of five different kinds of space based on these formalisms.

The first, *physical space* – the space of existence – was defined by the first four axioms. Physical space considers object displacement as mathematical vectors, combined with the concepts of physical mass and time. *Sensorimotor* space was defined by the first three axioms, combining existence and physical interaction at the scale of the human body. *Perceptual space*, which was defined by the first two axioms, link existence and physical interaction to salient environmental cues. *Cognitive spaces* link salient environmental cues to the cognitive factors of beliefs, knowledge, and memory. Due to the role of memory, cognitive space cannot be characterized by any of the five axioms – it is free of the constraints of physical space. This freedom from formal constraints is the critical distinction between perceptual and cognitive space. The last kind of space, *symbolic space*, was the space of maps and other kinds of symbolic (both tangible and intangible) representations of space, and involves the attachment of meaning with cognitive space.

Empirical evidence for manipulable object space

A body of literature has developed over the past ten or so years that offers evidence for manipulable object space, or the space within arm's reach of the body. Lang (1989), a linguist, proposed the Primary Perceptual Space (PPS) model to explain how physical space close to the body is 'reconstructed in terms of categorised sensory input from our senses' (p. 11). The PPS defines three distinct axes that define our internal model of space around the body. The first axis, the vertical axis, is defined by gravity and therefore provides the up–down schema. The vertical axis is constant and ubiquitous, and is 'superior' to the other two axes. The second axis, the observer axis, arises from sight, has an anatomically determined pivot allowing for a 180° panning angle, and therefore is defined by what we can see (front), and what we cannot see (back). The last axis, the horizontal axis, gives rise to left and right schema. This axis is not identified by primary perceptual information, but instead by the left and right symmetry of the body. A number of studies have explored the relevancy of Lang's three body axes to spatial cognitive representations of the space around the body (Tversky *et al.*, 1999).

Work by Franklin and Tversky (1990) and Bryant *et al.* (1992) demonstrated that people do construct a representation that in fact mirrors Lang's three axes. They termed this the *spatial framework model*. In this series of studies, subjects read a description of a space (e.g., an auditorium) which described the position of objects that were around the subject. When asked to respond to object probes, recall of objects placed around the body was fastest with respect to the head–feet axis, next fastest with regard to the front–back axis, and least accessible with respect to the left–right axis. Furthermore, Bryant *et al.* (1992) found that the spatial framework model was valid when subjects assumed a perspective from within a space, but not when the subjects assumed a viewpoint external to the space. In other words, the spatial framework model pertains to spaces that surround the body, not spaces experienced as if one were an outside observer. Additional evidence for this space was provided when subjects were reclined from an upright position. When reclined, none of the body axes were 'most salient', and recall of objects was made with respect to the body and not to the three axes. This research demonstrates that space within arm's reach of the body is a distinct space that is represented cognitively with the spatial framework model.

Empirical evidence for environmental space

Environmental space, spaces learned via navigation over time, has perhaps gathered more attention than any other type of space. Siegel and White (1975) first detailed a model for the learning of environmental spaces, a model that has now been termed the dominant framework (Montello, 1993).

In this model, space is represented by landmarks, routes and configurations (Freundschuh, 1991). When learning an environmental space, the dominant framework posits that landmarks are learned first, that routes connecting landmarks are learned subsequently, and finally configurations (distance and direction) of landmarks are learned last (Siegel and White, 1975). This learning takes place over time. Current thinking is that people do not learn an environment in the progression of landmarks, routes, and configurations, but that they are acquiring all three types of information continuously, refining the accuracy and precision over time (Montello, 1993).

What is generally accepted is that environmental space is schematized as landmarks (nodes), routes (links) and configurations (spatial relationships between features) (Freundschuh, 1991, 1992a; Montello, 1993), and that this information contains distortions with regard to distance (Couclelis *et al.*, 1987; Freundschuh *et al.*, 1990) and direction (Tversky, 1981). Couclelis *et al.* (1987) and Freundschuh *et al.* (1990) report studies showing an asymmetry in distances between landmarks of different salience. Couclelis *et al.* (1987) found that distance estimates toward an important landmark are less than distances estimated away from an important landmark. Freundschuh *et al.* (1990) report that distance estimates made while at a landmark were shorter for estimates from that landmark to another landmark, and longer from the distant landmark to the present landmark. Other studies show distance away from a city centre are judged shorter than distances toward a city centre, and landmarks separated by a line (e.g., a political boundary) on a map are judged to be further apart than those same landmarks not separated by a line.

Work by Lynch (1960) and Freundschuh (1991, 1992b) suggest that the layout of the road structure of an environment also impacts the nature of the cognitive representation. Lynch (1960) found that residents of Boston, which possesses a non-gridded road pattern, demonstrated greater distortions in distances and orientations than did residents of Los Angeles, which has a gridded road pattern. Residents of Boston also described the city as being confusing and difficult to navigate, whereas residents of Los Angeles deemed their city easy to navigate. Freundschuh (1992a, 1992b) found that older subjects (seventeen and eighteen years old), made more accurate distance and orientation estimates in a gridded environment than a non-gridded environment.

Empirical evidence for map space

Map spaces are representations of environmental and geographic spaces and, as such, offer spatial information that is either difficult, or not possible, to acquire first hand from these larger spaces. Maps show, at a glance, the relative locations of places and features in the real world. Large-scale maps (larger than 1:250,000) enable the map reader to see accurate distance and direction information, information that has been referred to in the literature as survey knowledge (Taylor and Tversky, 1992; Thorndyke and Hayes-Roth, 1982). Several studies have compared spatial knowledge acquired from maps versus navigation experience.

Thorndyke and Hayes-Roth (1982), in a study comparing spatial knowledge learned from maps (map learners) to spatial knowledge learned from first hand navigation experience (navigation learners) found several interesting differences. First, they found that map learners' estimates of Euclidean distances were more accurate than Euclidean distance estimates of navigation learners, and that route distance estimates for navigation learners were more accurate than route distance estimates of maps learners. Second, they found that map learners judged object location more accurately than did navigation learners, and that navigation learners made orientation estimates between landmarks more accurately than did map learners. The map learners, it is suggested, develop survey knowledge of an environmental space, whereas the navigation learners develop procedural knowledge (Thorndyke and Hayes-Roth, 1982).

Taylor and Tversky (1992), using descriptions of environmental spaces, found similar effects with route and survey descriptions. In four different experiments, Taylor and Tversky (1992) had subjects read route or survey descriptions of naturalistic environments, after which they answered verbatim or inference questions from both perspectives, and drew maps of the environments. The results of their study demonstrated that subjects who learned about an environmental space from a route description created a cognitive representation that supported recall of procedural knowledge, and subjects who learned from a survey description created a cognitive representation that supported recall of configurational knowledge.

Effects of size of space

Reconstructing spatial configurations

There is a fairly rich literature on the effects of the absolute size of a space on performance of spatial tasks. Most of this work is developmental in nature, and shows that children typically perform better in relatively small spaces than in relatively large spaces. Research by Acredolo (1981) and by Acredolo and Boulter (1984), that required children to retrieve a set of keys dropped by an experimenter in either a large room, or a small-scale model of that room, found that children performed better in the small space than in the large space. Research by Siegel *et al.* (1979), that required five- and seven-year-olds to reconstruct, from memory, the configuration of objects learned from a model room, found that the five-year-olds, but not the seven-year-olds, performed much worse reconstructing the configuration in a large space than in a small space. These studies demonstrate that the absolute size of a space affects spatial performance.

Expressing spatial language

Recent work by Freundschuh and Blades (1997, 1998) compared the understanding of spatial concepts in a small-scale, table-top space to spatial
concept understanding in a large-scale landscape model space. Spatial
concepts (or locatives) are phrases or words that denote place location.
Examples in English include terms such as in, under, near, far, through,
front, and back (Landau and Jackendoff, 1993). Locatives are essential for
descriptions of space. It is impossible to describe the relationships between
places and objects without locatives. Understanding any spatial information (directions, maps, geographical descriptions, etc.) depends on an
appreciation of the spatial terms used in the description. Despite their
importance, the way that locatives are used and understood has hardly been
investigated, particularly in spaces larger than the size of a small table-top.

Previous empirical research into locative understanding has focused almost
exclusively on children, and the way that they first understand these spatial
concepts. An exception to this is the work of Mark and Egenhofer (1994)
on adults' understanding of the spatial relationships between lines (a road)
and polygons (a park). Johnston (1984; Johnston and Slobin, 1979) found
that children acquired the following locatives in the order listed: in, on,
under, next to, between, and back/front (see also Durkin, 1981; Bailystock
and Codd, 1987; Bremner and Idowu, 1987; Conner and Chapman, 1985;
Cox and Isard, 1990; Johnston, 1988). The ability to understand these
locatives is an early achievement with children as they have some appreciation of these terms by the age of about five years.

However, in previous studies the experimenter scoring each response as
correct or incorrect measured children's performance. This assessment procedure has two drawbacks. First, the assessment of accuracy is based on the
experimenter's own implicit assumption of correct use of the term, rather
than on the performance of a control group. Second, a dichotomous criterion ignores any developmental differences in the way that children use a
term. For example, Sowden and Blades (1996) found a direct relationship
between age and the distance between two objects when subjects three-,
four- and six-years-old were instructed to place one object 'near' another.

Finally, previous research on spatial concept understanding has tested
subjects in only one spatial context – that of a table-top space. Consequently,
little is known about locative use and understanding in other, more realistic spaces. Contextual factors are particularly important in the use of
locatives because few locatives have a precise definition, and some (for
example, near, next to, far, across) are predominantly context dependent.
Context dependency is related to geographic scale – for example, the location of a bicycle can be described as 'near a house', but it would not usually
make sense to describe the bike's location as 'near Los Angeles'. In the work
by Freundschuh and Blades (1997, 1998), they tested the following spatial
concepts:

in
near
behind
on
far
in front
next to
between
away
near and next to
close and next to
next to and far
far and away
in and across from
on and in front

In this study, children were asked to place objects, such as a car or person, in both a small-scale table-top space and in a large-scale, landscape model space. Subjects were asked to perform tasks of the form: *put the {object} {spatial concept} the {referent}*. Example questions included 'put this car near the school' and 'put this person far from the block'.

Freundschuh and Blades (1997, 1998) found developmental differences for *all* locatives tested in both the table-top and landscape models, except for the locatives in and on, for which there were no age differences. For the locatives 'in front', 'behind', 'between', 'in and across from', and 'on and in front', older participants responded correctly significantly more often than younger participants, with few significant differences between seven- and nine-year olds, and adults. For the locatives 'away from' and 'far', older participants placed objects significantly further from the referent than younger participants. For the locatives 'near' and 'next to', older partici-pants placed objects significantly closer to the referent than younger participants did. In tasks combining two locatives that made reference to distance from the referent, for example 'near' and 'far', older participants made significantly greater differentiation between the two locatives than did younger participants. For example, when participants were instructed to 'put this fire hydrant *near* the fire station and *next to* the mailbox', three-year-olds placed the hydrant between the fire station and the mailbox, whereas older participants put the hydrant increasingly closer to the mailbox and further away from the fire station.

In addition, nearly 70 per cent of the three-year-olds recognized and observed contextual constraints in the landscape model – constraints that confine cars to roads and parking lots, and confine people and animals to sidewalks, paths, parks, and park benches. By five years of age, children's responses were comparable to those of adults. This result suggests that geographic spaces have characteristics that are quite different than simple table-top spaces, and that people recognize these constraints at an early age.

A final important result was found when distances between object placement and the referent were analysed as a proportion of model size. Participants placed objects farther from the referent in the landscape model than in the table-top model. It is not clear if this result is due to the size difference between the two spaces, or if this is a real difference in how subjects understood the spatial concepts in the two spaces. Consequently, further testing is needed to explore this result.

Spatial concepts and image schemas

Frank (1996) suggested that different size spaces offer contrasting experiences with the world and the features in it, and that these experiences have guided the object/field representations of the world in GIS. He asserted that spatial image schemata are divided into those schemata that are used, prototypically, in a small-size spaces, and others that are used prototypically in large-size spaces (which often involve movement of the observer through space). An image schema is the result of 'recurrent pattern, shape, and regularity in' bodily experiences, therefore creating knowledge structures that enable humans to reason about their environment (Johnson, 1987, pp. 27–8). In other words, current knowledge (image schemas) is applied to new and novel situations, as well as to familiar situations in order to make sense and reason about the world. Johnson (1987) and Lakoff (1987) contend that spatial and temporal image schemata play a fundamental role in reasoning about the world, and therefore, are central to spatial cognition.

Frank suggested that the schemata of *container* (in/out), *object*, *link*, *up/down*, *left/right*, *before/behind*, *surface*, *support*, *part/whole*, and *contact* generally relate to small-scale spaces. Experience in small-scale space is primarily with, but not limited to moveable objects that have sharply defined boundaries where the properties of these objects (invariance, sharp boundaries, joined, stacked, put into/out of containers, etc.) can be described in terms of a universal algebra (Birkhoff and Lipson, 1970).

> From an experiential point of view, objects are a basic experience of small [size] space. Object identity and sharp boundaries, invariance of shape and properties under movement, etc. are the most salient characteristics of objects. They are significantly different from the properties found in large [size] spaces.
>
> (Frank, 1996)

In a large-size space, the typical experience is one of locomotion in an environment that is learned piecemeal, over time, through integration of various perspectives. The most salient operation is navigation. Places are recognized, but they lack definite boundaries, and are connected by paths. There are points along paths that are not sharply determined, and paths cross woods and fields (i.e., areas) that do not have clear boundaries. Experience in large-size space leads to the theory of topology. The spatial image schemata more typically relating to large-size spaces include *place*, *path*, *near/far*, *centre-periphery*, *through*, *across*, and *scale*. Frank (1996) notes that there is a tendency to use experiences in small-size space to structure experiences in large-size space. The work by Frank, built upon the work of Johnson (1987) and Lakoff (1987), offers a potential distinction between spatial concepts that inherently refer to spatial cognition of small-size spaces, and to spatial cognition of large-size spaces. This work provides insights and a possible mechanism for distinguishing, cognitively, between different kinds of spaces.

Understanding scale

Few issues in cognitive mapping are simultaneously as important, and as poorly understood, as those involving scale. Issues of scale affect almost everything that geographers and other spatial scientists do. For example, geographic information systems offer system users the capability of representing spatial phenomena at different scales. Though this capability for multiple representations exists, designers of GIS have not really considered issues of scale and its impact on the communication of spatial information. Montello (1998) posits that there is a lack of 'systematic research' that offers a solid foundation for understanding the cognition of scale and scale transformations. Consequently, very little is known regarding the cognitive factors that are associated with understanding various levels of scale.

Model spaces and their referent spaces

Research on the understanding of scale models and simple maps has grown in the past fifteen years, focusing primarily on children's understanding of maps and models and their referent spaces. Research by DeLoache (1987, 1989) has found that the ability to understand that a model represents a real world space develops by the age of three. In these studies, objects were hidden in a model space, and children were instructed to retrieve the hidden object in the referent space. By three years of age, children can perform this task quite successfully (see also work by Blades and Cooke, 1994; Dow and Pick, 1992). It is important to note that though the children in these studies can understand the correspondence between the model and the referent space, these studies tell us very little about the understanding of scale information. To find the hidden toy in these various studies, the child does not have to use scale information (e.g., distances, proportions, etc.) to find the toy. The child has only to match the 'chair' that the toy was hidden under in the model space to the 'chair' in the large space (Blades and Cooke,

1994; Perner, 1992). Research by Liben *et al.* (1982) demonstrated that when children had to use scale information their performance deteriorated significantly. In their study, Liben *et al.* (1982) had children and adults reconstruct, in a classroom, the configuration of furniture learned previously. Some subjects learned the layout of the furniture from a scale model, and others learned the locations of the objects in the actual room. Subjects then reconstructed the room at the same size that they learned it, or at the other size. The children in this study could reconstruct the configuration of the furniture quite successfully in a space that was the same size of the learning space, but performance deteriorated significantly if subjects were asked to reconstruct the configuration in the other size space. The results of Liben *et al.* (1982) were similar to those demonstrated in a study by Herman and Siegel (1978). Research on children's understanding and use of simple scale models suggests that young children develop an understanding of the basic, symbolic correspondence between maps and models, and their referent spaces at around age three. Research also suggests that increasingly advanced understanding of scale properties between models and their referent spaces is expressed with increasing age (Liben and Yekel, 1996).

Aerial photographs as map space

During the past thirty years, there has been a substantial amount of research exploring children's abilities to use and understand aerial photographs (see Blaut, 1997 and Downs and Liben, 1997a, 1997b). In these studies, researchers asked children to identify features, perform simple navigation tasks between features, and to colour-code features shown on aerial photographs. Blaut and Stea (1971, 1974; see also Blaut, 1997) contend that children as young as three years of age can understand the veridical perspective shown on maps, that they understand that aerial photographs represent real world spaces, and that they can use aerial photographs to solve simple spatial problems. In contrast, Downs *et al.* (1988; see also Liben and Downs, 1991) contend that while children can identify some features shown on an aerial photograph and can use these photographs to solve simple navigation problems, their understanding of aerial photographs is not fully developed, and possesses many distortions and errors in understanding. There is compelling research to support both sides of this debate (see Blades and Spencer, 1986, 1987, 1990; DeLoache, 1987; Downs and Liben, 1987, 1988, 1989, 1991; Liben and Downs, 1992; Spencer and Blades, 1985; Spencer *et al.*, 1989).

Many of the errors that children make during aerial photograph interpretation involve confusions of scale. For example, a five-year-old claimed that a line on an aerial photograph that was a road, could not represent a road because the line was too narrow for a car. Another example occurred when a child claimed to see an individual fish in Lake Michigan, when the feature he was looking at was a boat. The opposing camps in this debate used aerial photographs of different scales. Blaut and Stea used large-scale images (1:5,000), whereas Downs *et al.*, used significantly smaller-scale images (1:12,000). Objects on a large-scale aerial photograph are easier to recognize and identify than those same objects on a smaller-scale photograph. Errors that children made in Downs *et al.*'s (1988) study suggest that the constraints of scale in aerial photographs and maps may be difficult to grasp. To understand the concept of scale, one must realize that scale constraints apply to all features on a map, and that people, cars, etc. will not be visible at a small scale (1:30,000), but will be visible at a large scale (1:3,000). The results of these studies indicate that while children can understand simple scale relationships shown on aerial photographs, more complex scale relations (i.e., smaller scale images) are difficult for them to comprehend. There is virtually no research that tests adults on these map reading abilities.

Directions for future research

Identifying types of space

Future research on multiple kinds of space is needed in a variety of areas. A critical area is studies designed to verify truly unique, differentiable spaces that result in different cognitive representations. Evidence from previous studies has been reviewed that illustrates the existence of manipulable space, environmental space, and map space. What is lacking is strong evidence that the other three space types – non-manipulable space, geographic space, and panoramic space – are different kinds of space. Though several studies mention some aspect of these spaces, little research has been conducted to verify that these spaces are unique, and result in different cognitive representations. In addition, studies are needed to explore the existence of more, or other kinds of spaces. For example, virtual reality (VR) technology is being used to explore the cognition of large-size, or environmental spaces. Do other sorts of spaces not yet identified exist?

Links between different types of space

Another area where research is needed is a comparison of the cognition of different spaces, and the applicability of extending the results of studies conducted in one space to the cognition of other spaces. A variety of studies exploring large-size space have utilized manipulable object representations of these spaces. For instance, studies have used maps (Lloyd, 1989), slides (Golledge *et al.*, 1993), 3-D models (Blades and Cooke, 1994), and computer simulations (Richardson and Montello, 1999), drawing conclusions and making general statements about spatial learning and cognitive spatial representations in general. Using the results of experiments conducted in one space, for example a manipulable object space, to study spatial learning, behaviour and representation of an environmental space may be problematic, as results from studying one space may not necessarily apply to another space. Understanding the links between different space types can help to realize the implications of research designed to explore one type of space, but conducted in a different space. An area of new research where this issue will certainly be relevant is the use of virtual spaces to study spatial cognition (see Chapter 7 of this book). Using virtual spaces in empirical studies on navigation and wayfinding, for example, is advantageous because it is more efficient with regard to human subject testing (both in time and money). Though advantageous, researchers realize that navigating in a virtual space does not provide the same sensory inputs as walking through an environment. Differences in the resulting cognitive representations from these two spaces need to be understood.

Spatial knowledge and types of space

A research need related to 'links between spaces' is the identification of the kinds of spatial knowledge acquired in each space. There have been a number of studies that have explored spatial knowledge from map space (Freundschuh, 1992a; Lloyd, 1989; MacEachren, 1992), from environmental space (Freundschuh, 1992a; Lloyd, 1989; Thorndyke and Hayes-Roth, 1982), and from manipulable object space (Blades and Cooke, 1994; Blades and Spencer, 1994; DeLoache, 1989). These studies have demonstrated that different kinds of spaces result in the acquisition of different kinds of spatial knowledge. For example, knowledge acquired from maps is survey in nature, and knowledge gained from first hand navigation experience is procedural (Thorndyke and Hayes-Roth, 1982). Studies designed to measure and compare spatial knowledge acquired from a variety of spaces would augment our knowledge of differences in space types, and assist in the differentiation between space types.

Scale effects

Another research need is the systematic study of scale effects on the cognition of spatial information (Montello and Golledge, 1999). We understand very little about the cognitive factors and abilities that are used to understand scale changes. How are scale transformations understood? How do people relate one size space to another size space – for instance, a map of the United States to the United States? What sorts of multiple representations are created to assist people in understanding the spatial relationship of their house to the market, to work, to a neighbouring city, to a distant state and country? What kinds of distortions develop in this process of understanding scale changes? Related to scale effects, studies are needed that test children through adulthood – from very young children to older adults (see Chapters 9 and 10 in this book)– on understanding scale over the life span. These types of longitudinal studies would provide a foundation for understanding the evolution of comprehension of scale.

Applications to GIS and geographic education

A final area where research is needed is the application of what we know about spatial cognition in different size and scale spaces to the development of geographic information systems, and the development of geographic education curriculums. One goal of GIS design is the creation of spatial information tools that result in effective communication of spatial information to the system user. These multiple conceptions of space need to be integrated into GIS in the form of multiple spatial models, presentations of spatial information, and dynamic user interfaces. 'In order to enable GIS users to interact with these simulated spaces as if they were interacting with the actual space, the GIS must present a world of spatial concepts that is as close as possible to the concepts used when reasoning about real world spaces' (Freundschuh and Egenhofer, 1997, p. 371). Research is needed that will take the knowledge gained in basic research and operationalize this knowledge in GIS – for instance, create a GIS that presents a user with a pictures of a region, a map of a region, and a virtual reality representation of the region.

Results of basic research on scale and size of spaces are useful in the development of new education curriculums in map use skills and geography. Maps are the primary mechanism by which geographers store geographic information. Maps, in the broadest sense of the definition, are representations of space. They include, but are not limited to, traditional paper maps, computer images, aerial photographs, satellite images, three-dimensional representations, cognitive maps, sketch maps, navigation directions, and digital files. Due to the complexity of maps, it is important for students to understand how to use maps and other spatial tools to learn, analyse and interpret spatial phenomenon. Understanding maps requires that map users have an understanding of scale. Studies are needed that provide baseline information on children's understanding of scale and scale concepts, and when understanding of certain concepts emerge.

References

Abler, R. (1987) What shall we say? To whom shall we speak? *Annals of the Association of American Geographers*, 77, 511–24.

Acredolo, L.P. (1981) Small- and large-scale spatial concepts in infancy and childhood. In Liben, L.S., Patterson, A.H. and Newcombe, N. (eds), *Spatial Representation and Behavior Across the Life Span: Theory and Application*, New York: Academic, pp. 63–81.

Acredolo, L.P. and Boulter, L.T. (1984) Effects of hierarchical organization on children's judgements of distance and direction. *Journal of Experimental Child Psychology*, 37, 409–25.

Bailystock, H. and Codd, Z. (1987) Children's interpretation of ambiguous spatial descriptions. *British Journal of Developmental Psychology*, 5, 205–11.

Birkhoff, G. and Lipson, J.D. (1970) Heterogeneous algebras. *Journal of Combinatorial Theory*, 8, 115–33.

Blades, M. and Cooke, Z. (1994) Young children's ability to understand a model as a spatial representation. *Journal of Genetic Psychology*, 155, 201–18.

—— and Spencer, C. (1986) Map use by young children. *Geography*, 71, 47–52.

—— and —— (1987) How do people use maps to navigate through the world? *Cartographica*, 24, 3, 64–75.

—— and —— (1990) The development of 3- to 6-year-olds' map using ability: The relative importance of landmarks and map alignment. *Journal of Genetic Psychology*, 151, 181–94.

—— and —— (1994) The development of children's ability to use spatial representations. *Advances in Child Development and Behavior* 25, 157–99.

Blaut, J.M. (1997) Children can. *Annals of the Association of American Geographers*, 87, 152–8.

—— (1999) Maps and spaces, *The Professional Geographer* 51, 510–15.

—— and Stea, D. (1971) Studies in geographic learning. *Annals of the Association of American Geographers*, 61, 387–93.

—— and —— (1974) Mapping at the age of three. *Journal of Geography*, 73, 5–9.

Bremner, G. and Idowu, T. (1987) Constructing favorable conditions for measuring the young child's understanding of the terms in, on and under. *International Journal of Behavioral Development*, 10, 89–98.

Bryant, D.J., Tversky, B. and Franklin, N. (1992) Internal and external spatial frameworks for representing described scenes. *Journal of Memory and Language*, 31, 74–98.

Canter, D. (1977) *The Psychology of Space*. New York: St Martin's Press.

Conner, P. and Chapman, R. (1985) The development of locative comprehension in Spanish. *Journal of Child Language*, 12, 109–23.

Couclelis, H. and Gale, N. (1986) Space and spaces. *Geografiska Annaler* 68, 1–12.

—— Golledge, R., Gale, N. and Tobler, W. (1987) Exploring the anchor-point hypothesis of spatial cognition. *Journal of Environmental Psychology*, 7, 99–122.

Cox, M.V. and Isard, S. (1990) Children's deictic and non-deictic interpretations of the spatial locatives 'In Front Of' and 'Behind'. *Journal of Child Language*, 17, 481–8.

DeLoache, J.S. (1987) Rapid change in the symbolic functioning of very young children, *Science*, 238, 1556–7.

—— (1989) The development of representation in young children. *Advances in Child Development and Behavior*, 22, 1–39.

Dow, G.A. and Pick, H.L. (1992) Young children's use of models and photographs as spatial representations. *Cognitive Development*, 7, 351–63.

Downs, R. and Liben, L. (1987) Children's understanding of maps. In Ellen, P. and Thinus-Blanc, C. (eds), *Cognitive Processes and Spatial Orientation in Animal and Man*. Boston, MA: Martinus-Nijhoff, 202–19.

—— and —— (1988) Through a map darkly: Understanding maps as spatial representations. *The Genetic Epistemologist*, 16, 11–18.

—— and —— (1989) Understanding maps as symbols: The development of map concepts in children. *Advances in Child Development and Behavior*, 22, 145–201.

—— and —— (1991) The development of expertise in geography: A cognitive-development approach to geographic education. *Annals of the Association of American Geographers*, 81, 304–27.

—— and —— (1997a) Geography and the development of spatial understanding. *Proceedings of The First Assessment: Research in Geographic Education*, The Gilbert M. Grosvenor Center for Geographic Education.

—— and —— (1997b) Piagetian pessimism and the mapping abilities of young children. *Annals of the Association of American Geographers*, 87, 168–77.

—— , —— and Daggs, D. (1988) On education and geographers: The role of cognitive development theory in geographic education. *Annals of the Association of American Geographers*, 78, 680–700.

—— and Stea, D. (1973) *Image and Environment*. Chicago, IL: Aldine.

—— and —— (1977) *Maps in Minds: Reflections on Cognitive Mapping*. New York: Harper and Row.

Durkin, K. (1981) Aspects of late language acquisition: school children's use and comprehension of prepositions. *First Language*, 2, 47–59.

Frank, A. (1996) The prevalence of objects with sharp boundaries in GIS. In Burrough, P. and Frank, A. (eds), *Geographic Objects with Indetermined Boundaries*. London: Taylor and Francis, pp. 29–40.

Franklin, N. and Tversky, B. (1990) Searching imagined environments. *Journal of Experimental Psychology: General*, 119, 63–76.

Freundschuh, S.M. (1991) The effect of the pattern of the environment on spatial knowledge acquisition. In Mark, D.M. and Frank, A. (eds), *Cognitive and Linguistic Aspects of Geographic Space*, Kluwer, pp. 167–83.

—— (1992a) Is there a relationship between spatial cognition and environmental patterns? In Frank, A.U., Campari, I. and Formentini, U. (eds), *Theories and Methods of Spatio-Temporal Reasoning in Geographic Space*, New York: Springer, pp. 288–304.

—— (1992b) *Spatial Knowledge Acquisition of Urban Environments from Maps and Navigation Experience*, PhD thesis, State University of New York at Buffalo.

—— and Blades, M. (1997) Locative understanding in large-scale (geographic) environments. Paper presented at the Annual Meeting of the Society for Research in Child Development, Washington, DC, 6–9 April.

—— and —— (1998) Put the horse NEXT to the lake and FAR from the water tower: The development of locative understanding in large- and small scale model spaces. Paper presented at the Annual Meeting of the Association of American Geographers, Boston, MA, 25–9 March.

—— and Egenhofer, M. (1997) Human conceptions of spaces: Implications for GIS, *Transactions in GIS*, 2, 361–75.

—— Mark, D.M., Gopal, S., Gould, M. and Couclelis, H. (1990) Verbal directions for wayfinding: Implications for navigation and geographic information and analysis systems. *Proceedings of the 4th International Symposium on Spatial Data Handling*, Zurich, Switzerland, pp. 478–87.

Gärling, T. and Golledge, R. (1993) *Behavior and Environment: Psychological and Geographical Approaches*. London: North Holland.

—— and —— (1997) Environmental perception and cognition. In Zube, E. and

Moore, G. (eds), *Advances in Environment, Behavior, and Design* (vol. 2). New York: Plenum: pp. 203–36

—— Ruggles, A., Pellegrino, J. and Gale, N. (1993) Integrating route knowledge in an unfamiliar neighborhood: Along and across route experiments. *Journal of Environmental Psychology* 13, 293–307.

Herman, J.F. and Siegel, A.W. (1978) The development of cognitive mapping of the large scale environment. *Journal of Experimental Child Psychology*, 26, 389–406.

Ittelson, W.H. (1973) Environment perception and contemporary perceptual theory. In Ittelson, W.H. (ed.), *Environment and Cognition*, New York: Seminar, pp. 1–19.

Johnson, M. (1987) *The Body in the Mind*. Chicago, IL: University of Chicago Press.

Johnston, J.R. (1984) Acquisition of locative meanings: behind and in front of. *Journal of Child Language*, 11, 407–22.

—— (1988) Children's verbal representation of spatial location. In Stiles-Davis, J., Kritchevsky, M. and Bellugi, U. (eds), *Spatial Cognition, Brain Bases and Development*. Hillsdale, NJ: Erlbaum, pp. 195–205.

Johnston, J. and Slobin, D. (1979) The development of locative expressions in English, Italian, Serbo-Croatian and Turkish. *Journal of Child Language*, 6, 529–45.

Kuipers, B. (1978) Modeling spatial knowledge. *Cognitive Science* 2, 129–53.

Lakoff, G. (1987) *Women, Fire, and Dangerous Things: What Categories Reveal About the Mind*. Chicago, IL: University of Chicago Press.

Landau, B. and Jackendoff, R. (1993) 'What' and 'where' in spatial language and cognition. *Behavioral and Brain Sciences*, 16, 217–65.

Lang, E. (1989) Natural language understanding and reference frames, in Mark, D.M., Frank, A.U., Egenhofer, M.J., Freundschuh, S.M., McGranaghan, M. and White, R.M., *Languages of Spatial Relations: Initiative Two Specialist Meeting Report*, NCGIA Technical Paper 89–2, 62 pp.

Liben, L.S. and Downs, R.M. (1989) Understanding maps as symbols: The development of map concepts in children. *Advances in Child Development and Behavior*, 22, 145–201.

—— and —— (1991) The role of graphic representations in understanding the world. In Downs, R.M., Liben, L.S. and Palermo, D.S. (eds), *Visions of Aesthetics, the Environment and Development: The Legacy of Joachim F. Wohlwill*. Hillsdale, NJ: Erlbaum, pp. 139–80.

—— and —— (1992) Developing and understanding of graphic representations in children and adults: The case of GEO-graphics. *Cognitive Development*, 7, 331–49.

—— , Moore, M.L. and Golbeck, S.L. (1982) Preschoolers' knowledge of their classroom environment: Evidence from small-scale and life-size spatial tasks. *Child Development*, 53, 1275–84.

—— and Yekel, C. (1996) Preschoolers' understanding of plan and oblique map: The role of geometric and representational correspondence. *Child Development*, 67, 2780–96.

Lloyd, R. (1989) Cognitive maps: Encoding and decoding information. *Annals of the Association of American Geographers*, 79, 101–24.

Lynch, K. (1960) *Image of The City*. Cambridge, MA: MIT.

MacEachren, A. (1992) Application of environmental learning theory to spatial knowledge acquisition from maps. *Annals of the Association of American Geographers* 82, 245–74

Mandler, J.M. (1983) Representation. In Mussen, P. (ed.), *Handbook of Child Psychology*, vol. III (4th edn), New York: Wiley, pp. 420–94.

Mark, D.M. and Egenhofer, M.J. (1994) Modeling spatial relations between lines and regions: combining formal mathematical models and human subjects testing. *Cartography and Geographic Information Systems*, 21, 195–212.

Montello, D. (1993) Scale and multiple psychologies of space. In Frank, A. and Campari, I. (eds), *Spatial Information Theory: A Theoretical Basis for GIS*. Lecture Notes in Computer Science 716, New York, Springer, pp. 312–21.

—— (1998) Scale and detail in the cognition of geographic information. http://www.ncgia.ucsb.edu/varenius/scale

—— and Freundschuh, S.M. (1995) Sources of spatial knowledge and their implications for GIS: An introduction, *Geographical Systems*, Special Issue on Spatial Cognitive Models, 2, 169–76.

—— and Golledge, R. (1999) *Scale and Detail in the Cognition of Geographic Space*, National Center for Geographic Information and Analysis, University of California, Santa Barbara, 75 pp.

Moore, G.T. and Golledge, R.T. (1976) *Environmental Knowing*. Stroudsberg, PA: Dowden, Hutchinson and Ross.

NCGIA (1988) Proposal to the Geography and Regional Science Program at the National Science Foundation, *NCGIA Technical Paper 89–2*, National Center for Geographic Information and Analysis, University of California, Santa Barbara.

Perner, J. (1992) *Understanding the Representational Mind*, Hillsdale, NJ: Erlbaum.

Richardson, A. and Montello, D. (1999) *Disorientation in Virtual Environments*, Paper presented at the Annual Meeting of the Association of American Geographers, Honolulu, Hawaii, 23–7 March.

Siegel, A.W. (1981) The externalization of cognitive maps by children and adults: In search of ways to ask better questions. In Liben, L.S., Patterson, A.H. and Newcombe, N. (eds), *Spatial Representation and Behavior Across the Life Span: Theory and Application*, New York: Academic, pp. 167–94.

—— Herman, J.F., Allen, G.L. and Kirasic, K. (1979) The development of cognitive maps of large and small scale spaces. *Child Development*, 50, 582–5.

—— and White, S.H. (1975) The development of spatial representations of large-scale environments. In Reese, H.W. (ed.), *Advances in Child Development and Behavior* (10), New York: Academic, 9–55.

Sowden, S. and Blades, M. (1996) Children's and adults' understanding of the locative prepositions 'next to' and 'near to'. *First Language*, 16, 287–99.

Spencer, C. and Blades, M. (1985) How children navigate. *Journal of Navigation*, 445–53.

—— , —— and Morsley, K. (1989) *The Child in the Physical Environment: The Development of Spatial Knowledge and Cognition*. Chichester: Wiley.

Taylor, H.A. and Tversky, B. (1992) Spatial mental models derived from survey and route descriptions. *Journal of Memory and Language*, 31, 261–92.

Thorndyke, P.W. and Hayes-Roth, B. (1982) Differences in spatial knowledge acquired from maps and navigation. *Cognitive Psychology* 12, 137–75

Tversky, B. (1981) Distortions in memory for maps. *Cognitive Psychology*, 13, 407–33.

—— Morrison, J.B., Franklin, N. and Bryant, D.J. (1999) Three spaces of spatial cognition, *The Professional Geographer* 51, 516–524.

Vasiliev, I., Freundschuh, S.M., Mark, D.M., Theisen, G.D. and McAvoy, J. (1990) What is a map? *The Cartographic Journal*, 27, 119–23.

Zubin, D. (1989) Oral presentation, NCGIA Initiative 2 Specialist Meeting, Santa Barbara, CA. Reported in Mark, D. (ed.), *Languages of Spatial Relations: Researchable Questions and NCGIA Research Agenda*, *NCGIA Report 89–2A*, NCGIA.

9 Cognitive mapping in childhood

David H. Uttal and Lisa S. Tan

Introduction

In the span of a few years, children go from being immobile to freely navigating in a host of environments. The developmental changes in mobility are accompanied by changes in children's ability to keep track of locations. Children must learn the layout of their homes, their neighbourhood, their schools, and many other environments. Most children learn all of these environments with apparent ease. However, that police in almost all urban districts devote substantial effort to finding lost children demonstrates the importance of children forming accurate cognitive maps (Cornell *et al.*, 1996).

The focus of this chapter is on the development of children's conceptions and mental representations of environments. Our chapter is organized like the others in this book; we review the past and present of cognitive mapping research and discuss possible directions for future work. However, before beginning our review, we discuss briefly our perspective on two themes that are of central importance in much research on the development of cognitive mapping: scale and representation.

Scale and the development of cognitive mapping

People possess knowledge of spaces of a variety of sizes or shapes, ranging from table-tops through continents. However, much of the research in spatial cognition has focused on relatively small-scale spaces, such as rooms or experimental laboratories (although there are important exceptions that are discussed in this chapter). One obvious reason for the focus on relatively small spaces is that it is very difficult to study children's knowledge of larger spaces. Children are exposed to large-scale spaces in numerous ways, and each child's knowledge of, and exposure to, the environment will vary. It is far easier, and more scientifically controlled, to investigate children's knowledge of small-scale environments that can be systematically controlled and manipulated. However, many researchers have challenged the focus of research on people's knowledge of relatively small-scale spaces. For example, behavioural geographers and environmental psychologists have stressed that the perceptual and cognitive processes that are used in large-scale space may differ fundamentally from those that are used in small-scale space (see Acredolo, 1981; Hart, 1979; Siegel and White, 1975; Montello, 1993; Montello and Golledge, 1999 for a discussion of these issues).

Because our focus is on children, we review work that has been conducted in a variety of different sized spaces. This is appropriate because the scale of space in which children navigate changes dramatically with development (e.g., Acredolo, 1981; Herman and Siegel, 1978; Weatherford, 1985). For example, the sizes of the spaces that toddlers know well are likely to be much smaller than those that elementary school children know well. Hence, most of the work on very young children has been conducted in what geographers would consider to be very small spaces. By the pre-school years, children begin to explore and know the environment beyond their homes, and consequently, the focus of research shifts from the home to the neighbourhood and school.

Representation

The second theme that plays a prominent role in much research on the development of cognitive mapping is *representation.* We use this term in two distinct ways. First, we refer extensively to *mental* representations of space. By this we mean how information about space is coded in the mind. For example, people may encode information in multiple ways – in terms of landmarks, routes, or a map-like survey of the environment. Most theories of the development of cognitive mapping have couched their work in terms of changes in the way in which children mentally represent spatial information.

The second sense of representation is *external,* symbolic representations of space, such as maps and scale models. Much of the information that people know about very large-scale environments would be difficult, if not impossible, to acquire from direct experience navigating in the world. For example, we could not easily learn the locations of several cities in Europe without looking at a map. Acquisition of knowledge of the large-scale environment therefore requires that people understand and use the information that maps can provide. We therefore have included discussion of the development of children's understanding of maps and models.

The past

Research on the development of cognitive mapping began in the early years of the twentieth century, although there were not consistent programmes of research until the late 1960s and early 1970s. Trowbridge (1913) conducted the earliest research on the development of cognitive mapping in childhood. He suggested that there were important similarities between the spatial strategies of young children and those of adults in 'primitive' (i.e., non-western) cultures for representing the large-scale environment. Both groups tended to represent the environment more in terms of relatively simple routes rather than in terms of integrated, survey-like maps. Although many of Trowbridge's ideas have since been discredited, his seminal work nevertheless was important because it foreshadowed the central theme of much research on the development of cognitive mapping: that changes in how children mentally represent large-scale space may be the mechanism of developmental change.

By the 1940s and 1950s, there were extensive studies of children's knowledge of both small and large-scale environments. Perhaps the most notable were those of Piaget (Piaget and Inhelder, 1956; Piaget *et al.*, 1960). Piaget claimed that there was a developmental progression in children's representation information. For example, the pre-operational child's mental representations of spatial relations were based solely on *topological* relations, which maintain only relations of grouping and order. A pre-operational child did not mentally represent spatial locations in terms of distance or angle.

Piaget also investigated children's knowledge of relatively large-scale spaces, such as the layout of their town. In these studies (Piaget *et al.*, 1960), the children were asked to make miniature models of the layout of their town. Piaget claimed that children's constructions of the layouts of their home area reflected how they mentally represented spatial information. For example, children younger than approximately six or seven tended to conflate Euclidean distance with other measures of similarity or interest. For example, one child placed a store that sold candies and toys much closer to his school than it was in actual Euclidean distance. Until approximately age nine, children's constructions failed to show a systematic integration of the locations into an organized form.

A noteworthy characteristic of Piaget's work was that he did not make a fundamental distinction between children's conceptions and representations of small- and large-scale environments. In both cases, the pre-operational child's representation of space was primarily topological. The child captured in his or her mental representation only the relative ordering of locations; he or she did not think about spatial relations in terms of the metric properties (distance and angle of the space).

Research on the development of cognitive mapping burgeoned in the late 1960s and early 1970s. Converging movements in geography and psychology led to heightened interest in how people, and particularly children, represent the large-scale physical environment (see Appleyard, 1970; Downs and Stea, 1977; Kosslyn *et al.*, 1974; Wohlwill, 1970). The most influential work within this tradition was Siegel and White's (1975) theory of the development of children's mental representations of large-scale environments. They proposed that children of different ages mentally represented locations in the environment in fundamentally different ways. Young children (pre-schoolers) tended to focus more on landmarks. For example, they might represent the location of a building only as 'near the school.' These landmark-based representations did not capture information about the spatial relations among the different landmarks.

By the latter pre-school years, children began to augment their representations to include linkages between landmarks; these linked representations are *route* representations. A route-based representation includes several locations, but it does not include information about the spatial relations among these locations. In essence, route-based representations are *ordinal*; they encode locations in an inmutable order but they do not encode the distance between these locations or the spatial relations among them.

The final stage in Siegel and White's theory of the development of cognitive mapping was *survey knowledge*. Survey knowledge encodes locations from an overhead or oblique view, and includes knowledge of the multiple spatial relations among multiple locations (Levine *et al.*, 1982). Survey knowledge involves abstracting one's thoughts about space from direct space; the knowledge, and its representation, is no longer tied to travel or finding one's way. Survey representations are akin to maps (Tversky, 1996), but this does not mean that the survey representation is a map in the head (Downs, 1981). Under some specific circumstances, survey knowledge allows the child to think abstractly about multiple relations and multiple landmarks, and to behave *as if* he or she had a 'map in the head', albeit with distortions and severe limitations.

Siegel and others tested this theory in many different contexts, including children's classrooms, neighbourhoods, and school grounds (Cousins *et al.*, 1983; Herman and Siegel, 1978; Siegel and Schadler, 1977). In general, their theory provided an adequate description of how children of different ages thought about large-scale space. However, subsequent research has revealed some limitations. For example, researchers have disagreed substantially on what is and is not a landmark (Presson and Montello, 1988; Newcombe, 1988) as well as on exactly what constitutes survey knowledge (Kitchin, 1996). Nevertheless, Siegel and White's work continues to stand out as the most comprehensive and generative theory of how children learn and mentally represent the large-scale environment.

Infancy

Another important development in the 1970s was the emergence of research on the development of cognitive mapping in the first two years of life. Much of the early research on the development of cognitive mapping in infancy focused on how very young children code spatial locations. Young infants tended to code locations in terms of egocentric relations, that is, in terms of their own bodies. Consequently, if an experimenter moved the infant, they would often not be able to keep track of the location of a hidden object. Older infants were more likely to use allocentric codings, which involve external reference frames that are not linked to the infants' own bodies (Acredolo, 1981 Bremner, 1978). Some studies suggested that the onset of allocentric coding was tied to the emergence of mobility. That is, the learning to crawl or creep affects children's representation of the environment. As children become mobile, they can no longer rely exclusively on their own bodies as the basis for coding locations because the relation between their bodies and the locations is constantly changing (see Bai and Bertenthal, 1992; Bremner, 1978; Bremner and Bryant, 1977).

Other studies investigated the emergence of detour behaviour in one- and two-year-olds. A fascinating demonstration in this regard was Rieser *et al.*'s (1982) investigation of the effects of exposure to the overhead view on very young children's ability to navigate a detour in a simple maze. The children (ages nine to twenty-five months) were asked to navigate through the maze to find their mothers. Some of the infants were first raised to chest height to provide an overall, aerial view of the maze and of their mother. The children were then placed on the ground inside the maze. The dependent variable was whether they would go around a barrier to reach their mother, who the infants could not see from the ground-level pers- pective. Exposure to the overhead view facilitated the 25-month-olds' performance. This study thus demonstrates the importance of thinking about space from an aerial perspective, even for very young children.

The present

Space constraints do not permit us to provide a detailed review of current research in cognitive mapping. In this section, we instead highlight some key findings of ongoing research to indicate what kinds of questions are being addressed. We have selected these issues because of their current influence in the field and their relation to the classic themes of represen- tational change that emerged in the early history of cognitive mapping research.

Infants and toddlers

Reflecting a general interest in infancy in cognitive development work, there is substantial current interest in the development of cognitive mapping in very young children. This work is especially important because it can help to shed light on the developmental origins of the abilities that have typically been studied in older children and adults. For example, Hermer and Spelke (1994, 1996) have conducted a series of studies on the emer- gence of the use of landmarks by very young children. The basic question that motivated this work was whether very young children (ages eighteen to twenty-four months) and adults would attend to the presence of a land- mark that was placed in one corner of a rectangular-shaped room. There was a panel in each corner of the room, behind which a toy could be hidden.

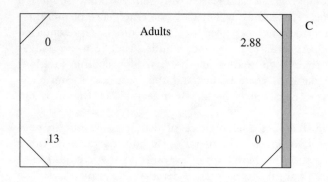

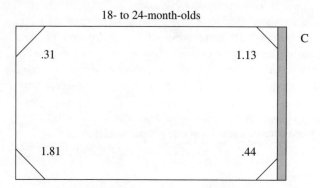

Figure 9.1 The design and results of Hermer and Spelke's (1996) studies. The numbers represent the average number of trials (out of four) on which subjects searched in each corner. The letter C represents the correct corner.

In the control conditions, all walls of the room were identical (other than their length). In the experimental conditions, one of the walls was covered entirely with blue fabric, as shown in Figure 9.1.

There were two important aspects of this design. First, because the room was rectangular, there were two sets of geometrically identical corners. That is, a short wall and a long wall connected in the same way in two sets of corners. Second, at least ostensibly, the presence of the blue curtain would seem to differentiate the two sets of corners. For example, although there were two corners in which a person could stand with a long wall in front and a short wall to the right, only one of the two corners would have had the blue curtain nearby. The critical question was whether children and adults would use the blue wall as a cue to differentiate the two sets of geometrically identical corners.

On each trial, participants in Hermer and Spelke's research observed as the toy was hidden in one of the four corners. Then, to disorient them, the participants were then asked to cover their eyes and turn around several times. The participant was then asked to uncover his or her eyes and to find the toy.

As shown in Figure 9.1, the young children did not use the landmark to differentiate the corners. They searched at the correct corner and at the geometrically equivalent corner almost equally, even though the blue curtain was placed near the correct corner. In contrast, adults were nearly perfect when the blue curtain was available, although they often searched at the geometrically identical corners when the blue curtain was not available. Follow-up studies showed that children could remember, point out, and describe the blue wall; nevertheless, they continued to fail to use it as a landmark to search for the toy.

Why would young children fail to use what would seem to be such an obvious landmark, especially when doing so would lead to near-perfect performance? Hermer and Spelke suggested a fascinating answer to this question: After they are disoriented by being turned several times, young children fall back to an evolutionary primitive strategy of relying on the shape of the environment. Previous work (e.g., Cheng, 1986; Gallistel, 1990) had shown that rats rely exclusively on the shape of the environment to find their way after disorientation. Hermer and Spelke suggested that the young children shared this evolutionary primitive cognitive 'module' (Fodor, 1983) that encodes *only* the shape of the environment. This cognitive module does not include any other information, such as the location or relevance of landmarks. In most situations, this strategy will work well; the shape of the environment is usually a reliable and stable cue that organisms can rely on when they are disoriented. However, in a perfectly rectangular space, children (and rats) cannot distinguish two of the four corners, and hence their searches are split equally between the correct corner and the geometrically identical corner. Development may therefore consist of learning when (and how) to ignore evolutionary primitive strategies in favour of using the cues that are best suited to the particular environment (Hermer and Spelke, 1996).

Other researchers are investigating the development of representation of distance information in infancy. For example, Bushnell *et al.* (1995) found that twelve-month-olds could find an object that was hidden under one of more than fifty irregularly shaped cushions in a large, circular, hiding space. Curtains were placed around the border of the search space and, consequently, there were no salient landmarks that the children could use as cues to the location of the object. The children were very good at finding the toy. These and similar results (e.g., Newcombe *et al.*, in press) have been difficult to explain without claiming that the children have accurately encoded (metric) distance information that specifies the location in terms of a specific distance (Bremner, 1993).

Maps and cognitive mapping

Until about fifteen years ago, almost all research on spatial cognition and cognitive mapping had focused on children's *mental* representations of spatial relations. However, in the past decade, researchers have devoted substantial attention to children's understanding of maps, scale models, and other external representations of space. For example, DeLoache and colleagues (DeLoache, in press, 1987, 1989, 1991, 1995; DeLoache and Burns, 1994; Marzolf and DeLoache, 1994; Uttal *et al.*, 1995) have focused on the emergence of the ability to use external representations (photographs, scale models, and simple maps) to find a hidden toy. In the basic version of this task, the child watches as the toy is hidden in the scale model (or as the location is indicated on a photograph). He or she is then asked to find the hidden toy in the corresponding location in the larger room.

In general, these studies have demonstrated that there is a dramatic improvement in children's ability to understand the relation between these external symbolic representations and the spaces that they represent. For example, in the standard scale model task (DeLoache, 1987), two-and-a-half-year-olds fail; their performance is at near-chance levels. In contrast, three-year-olds perform far better. These results have been interpreted (DeLoache, 1995; DeLoache and Smith, 1999) as indicating that the ability to understand external symbolic representations emerges sometime around the end of the third year. However, this does not mean that children of this age have a full understanding of the representational functions of maps. Studies that have investigated children's use of more complicated external representations indicate that the development of an understanding of maps and map-like representations continues well into the pre-school year. For example, when children look at 'real' maps, they often make errors that seem to reveal they do not fully grasp that the map is a *representation* of a geographic area. For example, one kindergartener said that a red line on a map (which indicated a road) could *not* represent a road because there are no red roads in the world (Liben, 1999). Additional experimental work supports some of these claims. For example, Liben and Yekel (1996) found that not until the latter elementary school years could children use the unique spatial position of an object on a map to disambiguate the locations of objects within their classroom (see also Blades and Cooke, 1994; Blades and Spencer, 1994; Liben and Downs, 1993). These results thus suggest that learning to use the *spatial* information that maps can provide may be a particularly difficult hurdle for young children. The question of how much young children know about the relations between maps and the spaces that maps represent continues to be a topic of substantial controversy (see Blaut, 1997; Downs and Liben, 1997; Liben, 1999; MacEachren, 1995).

Another line of research is investigating factors that may help children cope with the extra demands of using the spatial information available on maps. Since the time of the Gestaltists, psychologists have stressed that the interpretation of spatial stimuli is not merely a process of keeping track of individual locations. Instead, people often interpret individual locations as part of an organized or meaningful structure. A classic example is the constellations; ancient navigators organized and described the locations of sets of stars into meaningful patterns. This higher level of organization facilitates memory for and communication of locations. For example, it is easier to remember the location of a star if we can recall that it is in the handle of the Big Dipper.

Maps and map-like representations such as astronomical charts may play an important role in facilitating the construal of locations in terms of an organized or meaningful pattern. Because (most) maps represent spatial information from a plain or oblique perspective and at a relatively small scale, they can afford a fundamentally different way of perceiving and thinking about spatial information than can be gained from direct experience in the world. For example, it is easier to conceive of a set of stars as forming a constellation if we first see the locations on a chart.

In several recent studies, we (Uttal *et al.*, 1999) have investigated the development of the ability to use this function of maps. We have asked whether, and how, four- and five-year-old children can use maps to help them think about spatial information in ways that extend beyond the characteristics of individual locations. In this research, the children were asked to use a simple map to search for a sticker hidden under one of twenty-seven paper coasters that were distributed across the floor in the pattern shown in Figure 9.2. The overall configuration of objects formed the outline of a dog. Half of the children, the *lines* group, were informed that the locations could be interpreted as forming the outline of a dog; these children used the lines map shown in Figure 9.3. The remaining children, the *no lines* group, used the no lines map. There were no lines in the actual space

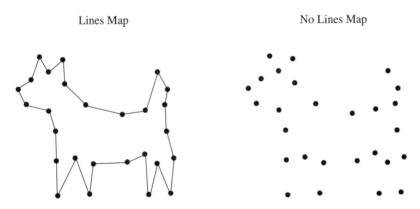

Lines Map   No Lines Map

Figure 9.2 The 'dog' figure used by Uttal *et al.* (1999).

Lines Map   No Lines Map

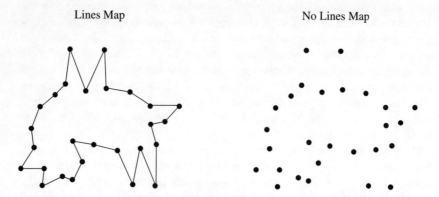

Figure 9.3 The alternate, meaningless figure used by Uttal *et al.* (1999).

in which the children searched. Thus the only difference between the two groups was the presence or absence of lines on the maps.

As predicted, the lines group performed substantially better than the no lines group. Analysis of children's errors revealed that knowledge of the dog pattern helped the children on difficult trials that required them to discriminate one search location from many others.

A follow-up study indicated that these effects could not be attributed simply to the addition of lines. We scrambled the dog to form the pattern shown in Figure 9.3. This alternate figure contained individual parts that were similar; however, the alternate figure lacked the systematic, overall organization of the dog figure. As in the original studies, half of the children saw a map on which the individual locations were connected with lines. In this case, adding lines to the figure made no difference in children's performance. These results suggest that it is the unique configuration of the locations – to form a meaningful pattern – which was responsible for the advantage that was observed in the first set of studies. Interestingly, however, we (Tan and Uttal, in preparation) have recently shown that children can perform well using the 'scrambled dog' pattern *if* they are first given prior experience in using the dog pattern. The children transferred what they learned about the dog pattern to the new pattern, which other children did not recognize as a meaningful shape. This result supports the claim that construing locations in terms of organized or meaningful patterns can be an important influence on the development of spatial cognition and map use.

Categorical representations of spatial locations

Another area of current interest concerns the development of *categorical* representations of space. When adults talk and think about the locations of objects in large-scale space, we often do so in terms of distinct categories (Allen, 1981; Huttenlocher *et al.*, 1994; McNamara, 1986; Sandberg *et al.*, 1996; Stevens and Coupe, 1978). We might say, for example, that Edinburgh is in Scotland or that Los Angeles is in California. These distinctions reflect a tendency of adults to break the world up into smaller units. We do this at many levels; we may think about particular locations within a neighbourhood, within a county, within a state, etc. Categorical representations are important because they allow us to know something about the location of a particular place without knowing the precise location. For example, if a person knows that a particular city is in California, then he or she knows some general information about the location of city, regardless of whether he or she knows the precise location. Of course, categorical representations can lead to systematic errors (McNamara, 1986; Stevens and Coupe, 1978), but they nevertheless provide important information in the absence of completely accurate information.

Several programmes of research are investigating the development of the ability to subdivide space into categories. For example, Huttenlocher *et al.* (1994) have investigated the emergence of the ability to subdivide a small-scale space into distinct regions or categories. Across several studies, Huttenlocher *et al.* tested very young children (ages sixteen to twenty-four months), pre-schoolers (ages four to five), first graders (age six to seven) and fifth graders (ages ten to eleven). The child's task was to search for a toy that was hidden in a five feet long, narrow sand box. The experimenter asked the child to turn around while an assistant hid a small toy in the sand. The child was then turned back to face the sand box and was allowed to search in the sand for the toy. A video recording was used to determine where the child searched in the sand.

The results revealed developmental differences in how younger and older children subdivided the sand box to remember the location of the toy. The youngest children (ages eighteen to twenty-four months) formed fewer spatial categories than did the older children and adults. This result was revealed in the pattern of children's errors. Specifically, the 18- to 24-month-olds' responses were biased toward the middle of the sand box. In other words, when the toy was hidden to the left of the centre of the sand box, these children tended to search somewhat to the right of the correct location. Likewise, when the toy was hidden to the right of the centre of the sand box, children's errors usually involved searching to the left. In contrast, the older children and adults demonstrated different patterns of bias. Their errors were biased towards the centre of the two *halves* of the sand box, suggesting that they had categorized the sand box as having two separate sections.

These results highlight the possibility of developmental change in the formation and use of spatial categories. An important aspect of development may involve learning to subdivide spaces into categories and to use those categories to facilitate memory (Huttenlocher *et al.*, 1994) communication (Plumert *et al.*, 1995), and map use (Acredolo and Boulter, 1984).

Studies of real world environments

Researchers are continuing to study the development of children's conceptions and mental representations of large-scale environments. One area of central interest concerns the development of flexible use of spatial information. Before approximately age twelve, children appear to have difficulty in choosing landmarks that will provide consistent and reliable cues to aid navigation. For example, twelve-year-olds are more likely to use distal and stable landmarks, such as tall buildings, as landmarks when learning the layout of a university campus (Cornell *et al.*, 1989; Cornell *et al.*, 1992; Heth *et al.*, 1997). Thus, even though young children can navigate successfully, children continue to fine tune their skills throughout the elementary school years (Allen and Kirasic, 1988; Pick and Lockman, 1981).

The work of Cornell, Heth and colleagues provides a good example of current approaches to the development of cognitive mapping in real world environments. Cornell *et al.* (1994) investigated how six-, and twelve-year-olds differed in their response to suggestions to use strategies to facilitate recall of landmarks while learning a new environment, the layout of a university campus. The experimenter led children on a route across the campus and then asked them to lead the way back. Different groups of children were given different instructions before they began the tour of the campus. Specifically, there were four conditions. In the *uninformed* condition, the children were not told that they would need to lead the way back. In the *generally informed* condition, the children were told only that they would be asked to lead the experimenter back to the starting point and that they should generally pay careful attention. Children in the third and fourth conditions were given more specific instructions that included information about the use of landmarks. In the *near landmark* condition, the experimenter stopped the children near two landmarks along the path and told them that the landmark might be useful for remembering the way back. For example, the experimenter pointed to a telephone booth and said, 'See this telephone booth? This telephone booth might be a good thing to remember for the way back' (p. 757). In the *far landmark* condition, the experimenter pointed to landmarks on the horizon rather than to landmarks on the path. The experimenter pointed to the tallest building on the skyline and said, 'See the tallest brick building? That's where we just came from.' The experimenter then turned and pointed to another building at the opposite end of the skyline and said, 'See that smokestack? That's where we are going?' (p. 757).

Children in all four conditions took a walk with an experimenter across the campus, following a standard route. At the end of the walk, the experimenter told the child that he or she would be the 'leader' for the return trip. The experimenters then assessed how far children travelled during the return trip, and how much of the total distance was spent on and off the original path. In general, the six-year-olds performed poorly, and they responded only to instructions to use the near landmarks. In contrast, the twelve-year-olds were able to use the distal landmarks to find their way back to the path when they deviated from it.

These results highlight the importance of learning to use the appropriate landmarks to keep track of one's location. Young children may be able to navigate successfully in environments with which they are familiar, but they have trouble keeping track of their location when they deviate from a known path. The results also have important practical locations for assessing and predicting the behaviour of lost children (Cornell *et al.*, 1996).

The future

In this final section, we consider what we believe to be the key issues that will receive attention in the coming decades. Some of these predictions reflect ongoing trends in research, and others reflect the likely influences of emerging technologies both on how children learn about environments and on the ability of researchers to study this development.

Relations between the cognition of small- and large-scale space

We believe that there are important similarities between the perceptual and cognitive processes that have been investigated in small-scale, laboratory tasks and those that exist in large-scale space. We foresee increasing collaboration between researchers who have studied small- and large-scale space. Accordingly, we expect to see an increasing number of studies that focus specifically on the relation between what is known about the development of cognition of small-scale space and the processes that develop in the cognition of large-scale space.

One example topic that could be investigated concerns the development of categorical representations of large-scale space. As discussed above, researchers (e.g., Huttenlocher *et al.*, 1994; Plumert *et al.*, 1995) have documented important developmental change in how children form spatial categories and use these categories to remember or communicate the location of objects. Almost all of these studies have been conducted in small-scale space, but we believe that the results may have important implications for research on the development of conceptions of large-scale environments. People often think about large-scale environments in terms of both formal (e.g., counties, states, countries, etc.) and informal (e.g., neighbourhoods, shopping districts, etc.) (Lynch, 1960). We know very little about the emergence of children's conceptions of these geographical scale spatial categories (although see Spencer *et al.*, 1989). A fruitful area of research would involve linking the emergence of children's categorical representations of small-scale space with their understanding of categorical representation in large-scale environment.

Emerging technologies and research on the development of cognitive mapping

We believe strongly that emerging technologies will play an increasingly important role in research on the development of cognitive mapping. We forecast a substantial increase in the use of two technologies that have recently become practical and inexpensive, virtual reality and electronic tracking systems. Each has substantial potential to contribute to research on the development of children's experience of large-scale environments.

Virtual reality technologies have received considerable attention in research on adult spatial cognition (e.g., Loomis *et al*., in press), but there has been relatively little virtual reality research with children. We expect this to change quickly. Virtual reality will allow researchers to expose children to many of the features of a realistic, large-scale environment while simultaneously maintaining experimental control. Researchers then can ask, and answer, many of the questions that have remained difficult to address in real world environments. For example, in a virtual environment, it should be possible to control precisely when and how children are exposed to different kinds of landmarks. This would allow the researcher to examine when, and how, children begin to integrate knowledge of these landmarks into survey-like representations.

Electronic tracking systems have been available for many years, but recent advances in global positioning systems (GPS) and related technologies has made them affordable. It is now relatively easy to keep track of a person's location at any moment. This technology could allow researchers to investigate how children travel in familiar environments and how they explore unfamiliar environments. This new data, in combination with appropriate sampling procedures (see Csikszentmihalyi, 1992) could be very useful in answering basic questions that thus far have received surprising little attention. For example, we know very little about how children's home ranges change with development, and whether there are sex differences in how boys and girls explore new environments (see Hart, 1979).

In sum, virtual environments and tracking technologies can facilitate research on the development of cognitive mapping in two related ways. First, it can allow us to know more about how children actually experience real environments, and second, it can allow us to simulate and control children's experience of environments. In combination, these two technologies will provide important insights into questions that previously were intractable.

What causes development?

Although research on cognitive mapping and spatial cognition has increased substantially in the past ten years, almost all research has focused on the *description* of developmental change. Very little work has attempted to explain how or why these changes occur. The focus on description of change rather than on explanation reflects the general focus of cognitive development work (see Siegler, 1996). However, in recent years, developmental scientists have begun to address in earnest the process and mechanisms of change. For example, developmental psychologists have studied the specific mechanisms of change in many domains, ranging from early motor development (Thelen and Smith, 1994) to the acquisition of strategies for solving mathematics problems (Siegler, 1996).

We predict that there will be an increase in research on the mechanisms of change in the development of cognitive mapping. A good example of the kind of work that we foresee concerns the influence of the onset of mobility on infants' representations of space (Bremner and Bryant, 1977; Bai and Bertenthal, 1992). This work illustrated that becoming mobile contributed to developmental change in children's ability to represent locations in space. We believe that there will be more studies of this type, involving intensive, short-term longitudinal (i.e., microgenetic) investigation of the process of change. One example concerns the acquisition of survey knowledge of large-scale environments. By testing children several times as they learn the layout of a new environment, researchers could gain insight into the process by which routes or landmarks are integrated into a more cohesive, integrated representation (see Uttal, 1999).

Conclusions

Although research on the development of cognitive mapping has grown and changed substantially in its eighty-year history, some core themes have remained constant throughout. For example, much of the research has focused, and will continue to focus, on how children's mental representations of the environment change with age. Although descriptions of both the form and content of these representations varies substantially from theory to theory, almost all researchers view the development of cognitive mapping as an interaction between the child's developing mental representations and the environments to which they are exposed.

Perhaps the most important question that researchers will face in the future concerns the nature of the environments that children will be asked to learn. It is important to note that the environments to which children are exposed will change dramatically in the future. So-called virtual environments will become as much a part of the child's everyday experience as the typical real world environments in which they navigate. Navigating on the World Wide Web or in other computer-mediated environments may become as important of an environmental experience as navigating in the home environment. Research will need to keep pace with changes in children's environments as they study how children's conceptions of these environments change.

Acknowledgements

Portions of this work were supported by Grant R29 HD 34929 from the National Institute for Child Health and Human Development and LIS Grant 97201313 from the National Science Foundation. We thank Rob Kitchin and Scott Freundschuh for their comments on this paper. Address correspondence to David Uttal, Department of Psychology, Northwestern University, 2029 Sheridan Road, Evanston, IL 60208–2710 USA.

References

Acredolo, L.P. (1981) Small- and large-scale spatial concepts in infancy and childhood. In Liben, L.S., Patterson, A.H. and Newcombe, N. (eds), *Spatial Representation and Behavior across the Lifespan*. New York: Academic, pp. 63–82.

—— and Boulter, L.T. (1984) Effects of hierarchical organization on children's judgement of distance and direction. *Journal of Experimental Child Psychology*, 37, 409–25.

Allen, G.L. (1981) A developmental perspective on the effects of 'subdividing' macrospatial experience. *Journal of Experimental Psychology: Human Learning and Memory*, 7, 120–32.

—— and Kirasic, K.C. (1988) Young children's spontaneous use of spatial frames of reference in a learning task. *British Journal of Developmental Psychology*, 6, 125–35.

Appleyard, D. (1970) Styles and methods of structuring a city. *Environment and Behavior*, 2, 100–17.

Bai, D.L. and Bertenthal, B.I. (1992) Locomotor status and the development of spatial search skills. *Child Development*, 63, 215–26.

Blades, M. and Cooke, Z. (1994) Young children's ability to understand a model as a spatial representation. *Journal of Genetic Psychology*, 155, 201–18.

—— and Spencer, C. (1994) The development of children's ability to use spatial representations. In Reese, H.W. (ed.), *Advances in Child Development and Behavior*, 25, San Diego, CA: Academic, pp. 157–99.

Blaut, J.M. (1997) Piagetian pessimism and the mapping abilities of young children: A rejoinder to Liben and Downs. *Annals of the Association of American Geographers*, 87, 168–77.

Bremner, J.G. (1978) Egocentric versus allocentric spatial coding in 9-month-old infants: Factors influencing the choice of code. *Developmental Psychology*, 14, 346–55.

—— (1993) Spatial representation in infancy and early childhood. In Pratt, C. and Garton, A.F. (eds), *Systems of Representation in Children: Development and Use*. Chichester: Wiley, pp. 211–34.

—— and Bryant, P.E. (1977) Place versus response as the basis of spatial errors made by young infants. *Journal of Experimental Child Psychology*, 23, 162–71.

Bushnell, E.W., McKenzie, B. E., Lawrence, D.A. and Connell, S. (1995) The spatial coding strategies of one-year-old infants in a locomotor search task. *Child Development*, 66, 937–58.

Cheng, K. (1986) A purely geometric module in the rat's spatial representation. *Cognition*, 23, 149–78.

Cornell, E.H, Heth, C.D. and Broda, L.S. (1989) Children's wayfinding: Response to instructions to use environmental landmarks. *Developmental Psychology*, 25, 755–64.

—— , —— , Alberts, D.M. Place recognition and wayfinding by children and adults, *Memory and Cognition*, 22, pp. 633–43.

—— , —— Kneubuhler, Y. and Sehgal, S. (1996) Serial position effects in children's route reversal errors: Implications for police search operations. *Applied Cognitive Psychology*, 10, 301–26.

—— , —— and Rowat, W.L. (1992) Wayfinding by children and adults: Response to instructions to use look-back and retrace strategies. *Developmental Psychology*, 28, 328–36.

Csikszentmihalyi, M. (1992) Validity and reliability of the Experience Sampling Method. In deVries, M. (ed.), *Investigating Mental Disorders in their Natural Settings*. Cambridge: Cambridge University Press, pp. 43–57.

Cousins, J.H., Siegel, A.W. and Maxwell, S.E. (1983) Wayfinding and cognitive mapping in large-scale environments: A test of a developmental model. *Journal of Experimental Child Psychology*, 35, 1–20.

DeLoache, J.S. (1987) Rapid change in the symbolic functioning of very young children. *Science*, 238, 1556–7.

—— (1989) Young children's understanding of the correspondence between a scale model and a larger space. *Cognitive Development*, 4, 121–39.

—— (1991) Symbolic functioning in very young children: Understanding of pictures and models. *Child Development*, 62, 736–52.

—— (1995) Early symbolic understanding and use. In Medin, D. (ed.), *The Psychology of Learning and Motivation*, 33. New York: Academic, pp. 65–114.

—— (in press). Dual representation and young children's use of scale models. *Child Development*.

—— and Burns, N.M. (1994) Early understanding of the representational function of pictures. *Cognition*, 52, 83–110.

—— and Smith, C.M. (1999) Early symbolic representation. In Sigel, I.E. (ed.), *Development of Mental Representation: Theories and Applications*. Mahwah, NJ: Erlbaum, pp. 61–86.

Downs, R.M. (1981) Maps and mappings as metaphors for spatial representations. In Liben, L.S., Patterson, A.H. and Newcombe, N. (eds), *Spatial Representation and Behavior across the Lifespan*. New York: Academic, pp. 143–66.

—— and Liben, L.S. (1997) The final summation: The defense rests. *Annals of the Association of American Geographers*, 87, 178–80.

—— and Stea, D. (1977) *Maps in Mind: Reflections on Cognitive Mapping*. New York: Harper and Row.

Fodor, J. (1983) *Modularity of Mind: An Essay on Faculty Psychology*. Cambridge, MA: MIT.

Gallistel, C.R. (1990) *The Organization of Learning*. Cambridge, MA: MIT.

Hart, R. (1979) *Children's Experience of Place*. New York: Irvington.

Herman, J.F. and Siegel, A.W. (1978) The development of cognitive mapping of the large-scale environment. *Journal of Experimental Child Psychology*, 26, 389–406.

Hermer, L. and Spelke, E. (1994) A geometric process for spatial reorientation in young children. *Nature*, 370, 57–9.

—— and —— (1996) Modularity and development: The case of spatial reorientation. *Cognition*, 61, 195–232.

Heth, C.D., Cornell, E.H. and Alberts, D.M. (1997) Differential use of landmarks by 8- and 12-year-old children during route reversal navigation. *Journal of Environmental Psychology*, 17, 199–213.

Huttenlocher, J., Newcombe, N. and Sandberg, E. (1994) The coding of spatial location in young children. *Cognitive Psychology*, 27, 115–48.

Kitchin, R.M. (1996) Methodological convergence in cognitive mapping research: Investigating configurational knowledge. *Journal of Environmental Psychology*, 16, 163–85.

Kosslyn, S.M., Pick, H.L. and Fariello, G.R. (1974) Cognitive maps in children and men. *Child Development*, 45, 707–16.

Levine, M., Jankovic, I.N. and Palij, M. (1982) Principle of spatial problem solving. *Journal of Experimental Psychology*, 111, 157–75.

Liben, L.S. (1999) Developing an understanding of external spatial representations. In Sigel, I.E. (ed.), *Development of Mental Representation: Theories and Applications*. Mahwah, NJ: Erlbaum, pp. 297–321.

—— and Downs, R.M. (1993) Understanding person-space-map relations: Cartographic and developmental perspectives. *Developmental Psychology*, 28, 739–52.

—— and Yekel, C.A. (1996) Preschoolers' understanding of plain and oblique maps: The role of geometric and representational correspondence. *Child Development*, 67, 2780–96.

Loomis, J.M., Blascovich, J.J. and Beall, A.C. (in press) Immersive virtual environments as a basic research tool in psychology. *Behavior Research Methods, Instruments, and Computers*.

Lynch, K. (1960) *The Image of the City*. Cambridge, MA: Technology.

MacEachren, A.M. (1995) *How Maps Work: Representation, Visualization and Design*. New York: Guilford Press.

McNamara, T.P. (1986) Mental representations of spatial relations. *Cognitive Psychology*, 18, 87–121.

Marzolf, D.P. and DeLoache, J.S. (1994) Transfer in young children's understanding of spatial representations. *Child Development*, 65, 1–15.

Montello, D.R. (1993) Scale and multiple psychologies of space. In Frank, A.U. and Campari, I. (eds), *Spatial Information Theory: A Theoretical Basis for GIS*. Berlin: Springer-Verlag, Lecture Notes in Computer Science 716, pp. 312–21.

—— and Golledge, R.G. (1999) Scale and detail in the cognition of geographic information. Report of Specialist Meeting of Project Varenius held 14–16 May 1998, in Santa Barbara, CA, University of California at Santa Barbara, http://www.ncgia.org.

Newcombe, N. (1988) The paradox of proximity in early spatial representation. *British Journal of Developmental Psychology*, 6, 376–8.

—— Huttenlocher, J., Learmonth, A. and Wiley, J. (in press) Spatial memory in 5-month-old infants. *Developmental Psychology*.

Piaget, J. and Inhelder, B. (1956) *The Child's Conception of Space*. New York: Norton.

—— , —— and Szeminska, A. (1960) *The Child's Conception of Geometry*. New York: Norton.

Pick, H.L. and Lockman, J.J. (1981) From frames of reference to spatial representations. In Liben, L.S., Patterson, A.H. and Newcombe, N. (eds), *Spatial Representation and Behavior across the Lifespan*. New York: Academic, pp. 39–62.

Plumert, J.M., Ewert, K. and Spear, S. (1995) The early development of children's communication about nested spatial relations. *Child Development*, 66, 959–69.

Presson, C.C. and Montello, D.R. (1988) Points of reference in spatial cognition: Stalking the elusive landmark. *British Journal of Developmental Psychology*, 6, 378–81.

Rieser, J.J., Doxsey, P.A., McCarrell, N.S. and Brooks, P.H. (1982) Wayfinding and toddlers' use of information from an aerial view of a maze. *Developmental Psychology*, 18, 714–20.

Sandberg, E.H., Huttenlocher, J. and Newcombe, N. (1996) The development of hierarchical representation of two-dimensional space. *Child Development*, 67, 721–39.

Siegel, A.W. and Schadler, M. (1977) Young children's cognitive maps of their classroom. *Child Development*, 48, 388–94.

—— and White, S.H. (1975) The development of spatial representations of large-scale environments. In Reese, H. (ed.), *Advances in Child Development and Behavior*, 10, New York: Academic, pp. 9–55.

Siegler, R.S. (1996) *Emerging Minds: The Process of Change in Children's Thinking.* New York: Oxford University Press.

Spencer, C., Blades, M. and Morsley, K. (1989) *The Child in the Physical Environment: The Development of Spatial Knowledge and Cognition.* Chichester: Wiley.

Stevens, A. and Coupe, P. (1978) Distortions in judged spatial relations. *Cognitive Psychology*, 10, 422–37.

Tan, L.S. and Uttal, D.H. (in preparation). From dog to hedgehog: Transfer in children's use of a meaningful patterns for maps.

Thelen, E. and Smith, L.B. (1994) *A Dynamic Systems Approach to the Development of Cognition and Action.* Cambridge, MA: MIT/Bradford Books.

Trowbridge, C.C. (1913) Fundamental methods of orientation and imaginary maps. *Science*, 38, 888–97.

Tversky, B. (1996) Spatial perspective in descriptions. In Bloom, P., Peterson, M.A., Nadel, L. and Garrett, M.F. (eds), *Language and Space: Language, Speech and Communication*, Cambridge, MA: MIT/Bradford Books, pp. 463–91.

Uttal, D.H. (1999) Seeing the big picture. Map use and the development of spatial cognition. Unpublished manuscript.

—— Gregg, V., Tan, L.S., Chamberlin, M. and Sines, A. (1999) Connecting the dots: Children's use of a meaningful pattern to facilitate mapping and search. Unpublished Manuscript.

—— Schreiber, J.C. and DeLoache, J.S. (1995) Waiting to use a symbol: The effects of delay on children's use of models. *Child Development*, 66, 1875–89.

Weatherford, D.L. (1985) Representing and manipulating spatial information from different environments: Models to neighborhoods. In Cohen, R. (ed.), *The Development of Spatial Cognition.* Hillsdale, NJ: Erlbaum, pp. 41–70.

Wohlwill, J.F. (1970) The emerging discipline of environmental psychology. *American Psychologist*, 25, 303–12.

10 Ageing and spatial behaviour in the elderly adult

K.C. Kirasic

Introduction

Is research on ageing and spatial behaviour in large-scale environments a worthwhile enterprise? A survey of projects funded by granting agencies, publications in leading journals in the cognitive and behavioural sciences, and presentations at learned societies in those disciplines over the past decade would unequivocally lead to the impression that it is not. For the most part, research on ageing and spatial cognition has been limited to a very few experimental studies examining the effect of declining information-processing speed on visual–spatial abilities and a partial handful of psychometric studies focusing on age-related stability or decline across different ability domains over the course of adulthood. Studies of actual wayfinding and orientation behaviour in older adults have been extremely rare.

Thus, as research disciplines, cognitive and behavioural psychology have indicated that this area of inquiry is insignificant. Certainly, this lack of significance is not due to a declining elderly population or to the insignificance of large-scale spatial behaviour in that population. Currently, about 13 per cent of the population are age sixty-five or over, and this segment is expected to grow to about 20 per cent in the next thirty years. Approximately, 84 per cent of individuals over sixty-five years of age in the United States currently hold a driver's licence, and they average 2.4 trips per day covering an average of 10.6 miles. It is anticipated that that nearly 94 per cent of 'Baby Boomers' will be active drivers when they reach age sixty-five, and they will be using their cars more frequently and for greater daily travel distances (Spain, 1997). This is ample motivation for the study of driving competence, but it must also be acknowledged that these drivers are not driving solely to exercise their perceptual-motor skills. While they drive, they navigate to their destinations. Thus, it is important then to have an understanding of spatial navigation and wayfinding abilities of the older adult driver.

Beyond driving, however, are the day to day spatial requirements of life. A variety of spatial skills are pressed into service as older adults find their way around in public buildings, use city and building maps, plan journeys on public transportation, and simply remain oriented during walks through familiar and unfamiliar territory. In short, independent travel is an important aspect of everyday life, and wayfinding and orientation skills are critically important for independent travel.

Regardless of how many older adults are in the population and how important wayfinding and orientation skills are for normal functioning in everyday life, it is possible that the consequences of older adults becoming disoriented in large-scale environments are not considered by the cognitive and behavioural science 'establishment' to be sufficiently severe to warrant serious study. In the absence of data, I resort to real world experiences to address this point.

In 1987, my sixty-eight-year-old mother, who resided in the suburban Pittsburgh area, called her older daughter, who lived nearby, in a somewhat confused state. She had been on a shopping expedition that included driving to three or four different stores in nearby communities, and when she returned to her car at the final destination, she realized that she did not know where she was. Of course, the store was familiar, but in that particular situation at that time, she could not establish the spatial relationship of the store to any familiar place. After conversing with my sister, my mother got a general impression of where she might be, and she eventually drove to an intersection where there were familiar landmarks. However, the episode was a shock, and she never felt as confident about her navigational skills after this episode as she did before.

In 1996, the seventy-two-year-old father of former graduate student at the University of South Carolina left his daughter's house in a North Dakota town to return to his home in Montana, a trip that he had made before. However, his daughter became concerned when he did not call to let her know that he had returned home safely. After many hours of worry, she finally received a call informing her that her father had missed his exit from the interstate highway and had travelled 500 miles beyond his destination. Although he returned home safely, neither he nor his family had confidence in his navigational skill after that time, and his independent travel efforts were curtailed.

In December 1997, my family and I were entering a shopping mall in the Columbia, South Carolina area on a rainy evening when we met a well-dressed woman approximately sixty years of age who was carrying a number of shopping bags and who was thoroughly soaked by the rain. She was upset, and with some hesitation she announced that she could not remember where she had parked her car. With a little assistance, she eventually realized that she had entered the mall through a different entrance from the one from which she was exiting. It is unknown if this episode of disorientation had long-lasting consequences, but it clearly resulted in acute confusion and loss of self-confidence within the specific situation.

Are these three situations representative of common events within society? If so, they suggest that older adults are at risk for disorientation in the context of a variety of large-scale spatial tasks and that the psychological and practical consequences of such disorientation are not trivial. The remainder of this chapter will include a brief review of work concerned specifically with ageing-related changes in wayfinding and orientation behaviour, an evaluative commentary on the current state of knowledge on this topic, and a look at the future of this research area. Throughout, the emphasis is on a functional perspective on the issues, never straying far from the purpose of trying to determine how cognitive and behavioural researchers can do work that can actually improvement quality of life for older individuals.

The past

In 1985, Kirasic and Allen provided a comprehensive review and critique of the status of spatial research as it pertained to the older adult (Kirasic and Allen, 1985). The psychometric, experimental, and ecological approaches served as the primary organizers of this literature. In the final analysis, it was concluded that the general view of an age-related decrement in the speed and efficiency with which a wide variety of spatial tasks are performed could not be accepted without qualification. Furthermore, it was put forth that the decrements observed in psychometric and traditional experimental investigations were more likely to be observed in tasks involving abstract components and unfamiliar contexts. It was suggested that tasks involving concrete components and familiar settings would not yield such findings. It was highlighted that the conclusions from the majority of these studies were severely limited primarily because very few studies have included consideration of real life spatial situations.

Kirasic and Allen (1985) concluded with a question and a challenge. The question had to do with the future directions in the study of ageing, spatial performance, and spatial competence. The challenge was for researchers to not only supply answers to questions but more importantly, to ask better questions. It was suggested that in order to ask those better questions, research in this area should consider three classes of phenomena: individual characteristics, adaptive processes, and spatial situations. Falling under the heading of individual characteristics were information processing abilities, personality variables, physical abilities, and neurological states. Adaptive processes included all the cognitive activities necessary for the performance of spatial tasks. Situations would include at the most general level the tasks and settings involving spatial behaviour. Given the complex daily agenda of the millions of non-institutionalized elderly adults, the opportunities for the study of spatial performance and spatial competence were numerous. It was hoped that this three-dimensional framework would excite a new and progressive phase of spatial investigation.

The present

How has this field of study responded to the above question and challenge? A review of the literature over the past fifteen years shows with few exceptions a striking lack of empirical efforts that could even come close to being called an investigation of spatial behaviour in the older adult. It appears that the 'Decade of the Brain' has cast its shadow over the possibility of any progress in the study of spatial behaviour. Large strides have been made in the study of brain mechanisms involved in object perception and spatial attention because of data collected from stroke victims, young and old. But ageing is not equivalent to having strokes, and the fact that some stroke victims show hemi-field neglect tells us little about the spatial behaviour of normal older adults.

Psychometric-type measures of spatial ability continue to be employed by researchers in psychology. Two fundamental lines of research underlie their use. One line of research focuses on the processing resources associated with cognitive performance, namely speed of processing and working memory. Typically, these studies show an age-related decline in spatial abilities across the board, with much of this decline attributed to decreased speed of information processing. Results such as these raise the issue of whether processing speed is chiefly responsible for age-related differences found in large-scale spatial tasks. It may be the case that older adults are at risk for spatial disorientation because of a decline in generic processing resources, but the empirical studies necessary to examine this idea have not been done.

The other path taken in the psychometric investigation of spatial abilities is one that attempts to localize the spatial processing areas of the human brain. Neural imaging, cerebral blood flow measures, and PET scans are commonly used technologies to investigate spatial visualization, spatial perception, and spatial orientation abilities. While admirable in their mission, these researchers have little to offer the investigator of spatial behaviour. This statement must be qualified however. These findings have direct relevance to therapists providing rehabilitation to elderly adults who have suffered any cerebral injury. But again, what can be taken to apply to the active independent older individual is at best speculative and at worse vague.

Fortunately, there has been some empirical research dealing directly with spatial cognition and behaviour in large-scale environments. This work can be differentiated into studies conducted exclusively in the lab and studies that included data collection in real world environmentts.

Learning routes and spatial layouts in the lab

Critical to navigating in general is the ability to maintain in memory some general sense of the spatial and temporal relations between points along a route or the relationship of adjacent locations. In an attempt to identify changes in the ability to learn and employ critical route events (landmarks) in describing and drawing the routes portrayed via slide presentation, Lipman (1991) used a procedure that required subjects to view a two route walk either one or three times. Subsequently, participants reported what they remembered from the walk and then sketched a map of the walk. Additionally, participants were asked to sort a randomly arranged set of photographs according to the route depicted in the scenes and to select the six most important scenes that would serve as aids to remembering the route.

Results indicated significant age differences in the content and organization of route memory. The older adults recalled significantly fewer landmarks than did the college students but not significantly fewer than did the adolescents or the middle aged participants. More important, however, was the finding that older adults organized the spatial information according to the distinctiveness of landmarks rather than their sequential order. This tendency on the part of older adults to rely on the non-spatial characteristics of routes may underlie their poorer performance observed in unfamiliar environments (Lipman, 1991).

Another study of route learning in younger and older adults was conducted in my laboratory, using a procedure in which a slide presentation was used to simulate a walk along a pre-selected route. In the first study, distance estimate data were collected to determine how accurately participants learned the spatial relations among features along the walk, and eye-movement data made it possible to determine what areas of the scenes attracted and held the attention of good and poor performers in the distance estimation task. Preliminary analysis of the distance estimate correlation scores indicated, as one would expect, that the older participants made significantly less accurate distance estimates than did their younger counterparts. However, it was noted that while 60 per cent of the elderly and 65 per cent of the young would be considered excellent estimators of distances ($r = 0.90$ or above), the older group had considerably more individuals scoring in the extremely poor range ($r = 0.60$ or lower) thereby lowering the overall performance mean of the older group.

With regard to the eye-movement data that had been collected, each scene was first divided into twelve sectors for analysis. The sector analysis indicated that, overall, participants scanned over fewer segments in high information scenes than in low information scenes. Furthermore, it was found that excellent estimators of distances looked at fewer sectors in both high and low information scenes than did poorer distance estimators. These findings suggest that high skill individuals, irrespective of age, had a better sense of what to look for − or where to look − in the presentation of route information. The lower performers' tendency to scan scenes more widely suggested a lack of this sense.

These results inspired another study under my direction, this one focusing on the possibility of training better performance in those participants who initially performed poorly in the distance estimation task. In this procedure, after a route was viewed twice and estimations were made, actual to estimate distance correlations were calculated for each participant. The groups were broken down into excellent, medium, and poor estimators. Individuals returning for their second testing session were placed into one of three training groups: the map group, verbal group, or no representation group. As the names indicated, the map group studied a map of the route, the verbal group heard a description, and the no representation group served as a control. After these treatments, all participants viewed a second slide presented route and made distance estimates to various features along that route (see Figure 10.1).

Analyses indicated that while excellent estimators were not influenced by the type of training group to which they were assigned, the performance of the medium and poor estimators were most markedly improved by viewing a map prior to the presentation of the walk. Verbal descriptions also led to improved distance estimation performance. Providing no information to the poor estimators showed itself to be the most deleterious to their performance. The reason underlying their poor performance may very well be explained by the Lipman and Caplan (1992) findings. The marked improvement in the verbal and particularly in the map condition suggests that spatial knowledge can be enhanced by a variety of supplemental information tools. Although age itself was not a factor in this training study, one may infer that supplementing or providing multiple sources of spatial information to not only older but also younger adults would positively impact their spatial behaviour. This should be especially true for the older adults who tend to show poorer performance on a variety of spatial tasks (Kirasic, 1999c).

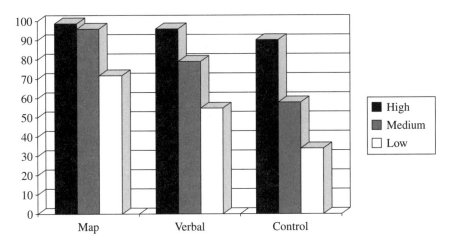

Figure 10.1 Percentage correct at three levels of expertise in different test conditions.

The potential benefit of providing multiple or supplemental forms of spatial information was part of the motivation for a study under my direction in which young and elderly adults produced written descriptions and sketch maps of a large-scale environments after being presented either a map or a verbal description of that environment. The design of the study called for all participants to produce both a written description and a sketch map; however, their expectations were manipulated so that one of the tasks came unexpectedly. Not surprisingly, all participants tended to perform better on the task that they expected to perform. However, older individuals were affected more than younger individuals, especially when they expected to describe the environment but were first asked, unexpectedly, to draw a map of the area. The results of this study suggest that maps and descriptions may be considered rather different (see Taylor, this volume), potentially complementary sources of spatial information; logically, two sources may be better than one when attempting to provide travellers with a sense of environmental layout prior to their experiencing that area. Also, the findings underscore the potential importance of compatibility between study tasks and actual performance in the environment (Kirasic, 1999d) (see Figures 10.2 and 10.3).

Many of the environments that adults, regardless of age, must deal with conform to certain design features (e.g., shopping malls, supermarkets), such that individuals may develop relatively articulated schemas for such places. Arbuckle *et al.* (1994) focused on potential schemas of this type in a study involving typical and atypical house floor plans. Participants studied these floor plans for the purpose of remembering the names assigned to different rooms in the floor plans. All participants did the task with

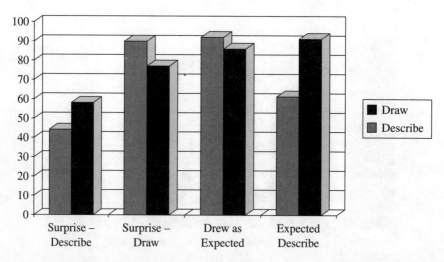

Figure 10.2 Performance differences in relation to test conditions (percentage correct) for the elderly adults.

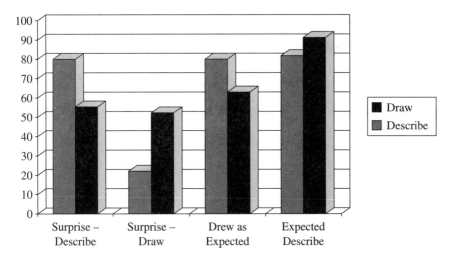

Figure 10.3 Performance differences in relation to test conditions (percentage correct) for the young adults.

both typical and atypical floor plans. When the typical floor plan was encountered first, the results showed that younger and older adults performed with equal accuracy when the layout of the house conformed with the typical layout. Results also indicated that older adults were much less accurate than younger adults when the layout of the house violated typicality. The older adults showed the same effect when the atypical floor plan was encountered first. The authors were hesitant to suggest any compensatory effect of schema-fitting environmental designs specifically for older adults. They would agree, however, that spatial configurations that do not approximate an expected typical configuration have a significant negative effect on performance.

The implication of all of these findings may be particularly relevant for urban designers and architects. Three conclusions can effectively be drawn from this work. First, by understanding the tendency of the older adult to employ non-spatial cues in the learning of anything from a route to an environment, more effort might go into providing stable non-spatial landmarks as referents in navigation. Secondly, multiple forms of representations to supplement each other (map and verbal description) would be particularly helpful in the learning of new spatial arrays. Finally, it would be important to keep in mind the potential importance in conforming to design expectations in order for the individual of any age to be able to utilize any pre-established schema to enhance wayfinding.

Studies in real-world environments

Is the study of spatial behaviour languishing in the laboratories of universities or are there a few valiant individuals attempting to explore actual spatial abilities and behaviours outside the laboratory? Gladly, there are a few. The work that has been done can be classified as falling under the heading of route learning, neighbourhood use, and environmental knowledge and wayfinding.

In two studies that investigated the learning of novel routes by younger and older adults, it was found that route learning ability declined with increasing age (Wilkniss *et al.*, 1997; Barrash, 1994). In the study by Wilkniss *et al.* (1997), which took place in a university medical centre, older adults had greater difficulty than younger adults in retracing the route and temporally ordering landmarks but were as good as younger adults in the recognition of landmarks occurring along the route. It was also indicated that the older adults had greater difficulty in memorizing and navigating the route after studying a two-dimensional representation of the to be learned route.

The results of Barrash's (1994) study, which also took place in a university medical school, indicated that the older adults made four times as many errors as the youngest group on the first trial and proved to be more striking over testing trials. It can be concluded that the findings are consistent with the results repeatedly found regarding the learning of novel routes or novel environments (Kirasic, 1989; Kirasic *et al.*, 1992; Kirasic, 1999a, Kirasic and Mathesis, 1990; Kirasic and Bernicki, 1990). In these studies, the older adult was at a significant disadvantage in learning new or novel environmental information. However, when tasks were presented in more familiar environments, age differences were negligible. It is apparent that familiarity is a very important determinant of successful wayfinding in the day to day world. Consistent with this view, an analysis of travel diaries of older adults indicated that older adults tended to take more familiar routes in their everyday travel (Kirasic, 1999b).

In what continues to be one of the most comprehensive investigations of spatial ability and knowledge, Simon *et al.* (1992) further analysed their archival data to determine if laboratory tests of spatial cognition were significant predictors of older adults' use of their neighbourhoods. The analyses indicated first that the laboratory-based measures of spatial cognition and subjects' overall knowledge of their neighbourhoods were both significant predictors of their use of goods and services in their neighbourhoods. Second, when contextual cues were not provided, the ability to learn and remember the location and orientation of objects were more predictive of neighbourhood use than any other measure reflecting familiarity with the area. Finally, in an attempt to model the predictors of neighbourhood spatial behaviour, the researchers found that The Building Memory test, which is a marker test for visual memory ability, successfully predicted neighbourhood knowledge, which was predictive of neighbourhood use.

A study conducted under my direction attempted to provide a model that best describes the relationship occurring among age, general spatial ability as assessed by psychometric tests, macro-spatial skills as assessed by experimental laboratory tasks, environmental knowledge as assessed in a field experiment, and wayfinding behaviour as observed in the field, in this case, a supermarket (Kirasic, 1999a). In this study, a global, all-encompassing spatial factor was found to be a significant mediator of the relationship between age and the acquisition of knowledge about environmental layout. It was suggested that future studies focus on developing a clearer delineation of spatial abilities and their relation to macro-spatial skills, and to identify a common mechanism or process that is tapped by psychometric tests of spatial ability.

A final study to be described comes from a different perspective on spatial cognition and behaviour and offers an interesting and promising approach for future investigation. Bhalla and Proffitt (in press) tested younger and older adults in a task in which participants were asked to estimate the geographic slant of hills in the environment. Typically, participants of all ages overestimate the geographic slant of hills using verbal judgements but reveal very accurate estimates using motor responses. Results of the study involved older adults showed that older adults overestimated slant to a greater extent than did younger adults in terms of verbal estimates, but that these estimates were reliably related to motorically derived estimates of slant, as was the case with all participants. These findings were taken as evidence that older adults recalibrate the link between visually guided action, which is not mediated by awareness or deliberation, and visually based judgements, which are grounded in awareness and intentionality. These findings have important implications for research on ageing and spatial behaviour. Specifically, they suggest that distance estimates and other judgements of spatial extent that are produced in the majority of experiments show systematic age-related distortion consistent with the effort that would be required to interact with the spatial area physically. Obviously, this finding points to the need for careful studies examining this effect and how the calibration between perception–action systems on the one hand and cognitive systems of representation on the other hand changes over late adulthood.

The future

It is surprising that so few studies have attempted to pursue understanding regarding the spatial learning and behaviour of the older adult. On the negative side, one can lament the lack of interest shown toward this research area. However, on the positive side, one can say that the field is ripe for any innovative, creative and curious approaches to this topic. This field is open and waiting for clever and dedicated researchers who are aware of the implications of an ever-growing population of elderly adults. It is predicted that this growing cohort of older adults will be healthier, more active, more mobile, and more demanding than any other aged group of past generations. They may continue to be fully employed, active in volunteerism, or the average individual could be out on a daily shopping trip or on a cross-continental trip. How can information that we provide make these environments more accessible?

The information being obtained in animal laboratories, in rehabilitation clinics with stroke and head-injured individuals, and in experimental psychology laboratories can provide the perceptive individual with valuable information in order to ask *better* questions. It is important not to view this type of information as an end in itself. The thoughtful researcher must step back in order to see how or if it fits in the bigger question of spatial behaviour exhibited by the elderly adult.

It is interesting to speculate on how new technologies may stimulate the study of ageing and spatial behaviour. One of the most often cited difficulties associated with the study of wayfinding and orientation in real world environments is the logistical challenge, i.e., getting researchers and participants together in an appropriate environment at a mutually satisfactory time in good weather and so forth. One technological advance that may alleviate some of these logistical challenges is the development of virtual environments. Using a relatively simple display, Nadel *et al.* (1998) have already demonstrated age-related declines in place learning in adults. As virtual environment technology becomes more widespread and the cost of displaying more complex environments decreases, it is easy to image this approach to studying spatial behaviour in young and old adults alike will yield empirical and conceptual benefits.

In closing, I wish to point to three areas of research that can be used as springboards to more innovative and meaningful questions. The first is continued investigation of spatial learning abilities across the life-span, the second is further investigation of acquired skill in wayfinding ability, and the third is a concerted effort to articulate environmental use and navigational patterns exhibited by the older adult in an attempt to make the environment more accessible. These three areas can easily be incorporated into the three-dimensional framework for research on spatial competence in elderly adults suggested by Kirasic and Allen (1985). All one need do is to look at the collection of studies that have been cited in this chapter. What has been presented is a patchwork quilt of studies, lacking a research agenda that could and should lead to the logical next question. If there is anything that researchers of macro-spatial abilities know, it is the importance of an interactionist perspective. Employing the proposed, or some other experimental framework that brings the interactive qualities of macro-spatial behaviour to light, is the future of research in this area.

If after fourteen more years little knowledge is gained regarding the spatial world of the elderly adult, we will need to ask why. Is there no more to learn? Have we run out of questions? Have we been seduced by technology that does not lend itself well to work with an older population? Will we all enter our later years and experience the lack of progress made by investigators in this area?

References

Arbuckle, T.Y., Cooney, R., Milne, J., and Melchoir, A. (1994) Memory for spatial layouts in relation to age and schema typicality. *Psychology and Aging*, 9, 467–80.

Barrash, J. (1994) Age-related decline in route learning ability. *Developmental Neuropsychology*, 10, 189–201.

Bhalla, M. and Proffitt, D.R. (in press) Visual-motor recalibration in geographical slant perception. *Journal of Experimental Psychology: Human Perception and Performance*.

Kirasic, K.C. (1989) The effects of age and environmental familiarity on adults' spatial problem solving performance: Evidence of a hometown advantage. *Experimental Aging Research*, 15, 181–7.

—— (1999a). Age-sensitive spatial abilities, knowledge of environmental layout, and wayfinding mediated relationships and missing links. Under review.

—— (1999b). Predictors of spatial perspective taking and perspective verification ability in young and elderly adults. In preparation.

—— (1999c). Expertise and attentional components in the processing of route information. In preparation.

—— (1999d). The utility of map vs verbal directions for elderly adults. Presented at the annual meeting of the Association of American Geographers, Honolulu, Hawaii.

—— and Allen, G.L. (1985) Aging, spatial performance and spatial competence. In Charness, N. (ed.), *Aging and Human Performance*. New York: Wiley.

—— , —— and Haggerty, D. (1992) Age-related differences in adults' macro-spatial cognitive processes. *Experimental Ageing Research*, 18, 33–9.

—— and Bernicki, M.B. (1990) Acquisition of spatial knowledge under conditions of temporal discontinuity in young and elderly adults. *Psychological Research*, 52, 76–9.

—— and Mathes, E.A. (1990) Effects of different means of conveying environmental information on elderly adults' spatial cognition and behavior. *Environment and Behavior*, 22, 591–607.

Lipman, P.D. (1991) Age and exposure differences in the acquisition of route information. *Psychology and Aging*, 6, 128–33.

—— and Caplan, L.J. (1992) Adult age differences in memory for routes: Effects of instruction and spatial diagram. *Psychology and Aging*, 7, 435–42.

Nadel, L., Thomas, K.G.F., Laurance, H.E., Skelton, R., Tal, T. and Jacobs, W.J. (1998) Human spatial cognition in a virtual arena. In Freksa, C., Habel, C. and Wender, K.F. (eds), *Spatial Cognition – An interdisciplinary approach to representation and processing of spatial knowledge*. Berlin: Springer-Verlag, pp. 399–427.

Simon, S.L., Walsh, D.A., Regnier, V.A. and Kraus, I.K. (1992) Spatial cognition and neighborhood use: The relationship in older adults. *Psychology and Aging*, 7, 389–94.

178   *K.C. Kirasic*

Spain, D. (1997). *Societal trends: The ageing baby boom and women's increased indepen-dence. Final report prepared for the Federal Highway Administration*, US Department of Transportation.

Wilkniss, S.M., Jones, M.G., Korol, D.L., Gold, P.E. and Manning, C.A. (1997) Age-related differences in an ecologically based study of route learning. *Psychology and Aging*, 12, 372–5.

11 A view of space through language

Holly A. Taylor

Introduction

Can language processing provide clues about spatial cognition? Samuel Taylor Coleridge (1772–1834) hinted at an answer to this question when he said, 'The best part of human language, properly so called, is derived from reflection on the acts of the mind itself' (Coleridge, 1834). On first reflection, the connection between spatial processing and language processing more often seems one of opposites. Laterality of function places spatial and linguistic processing predominantly in separate brain hemispheres. Individual differences generally favour either spatial or linguistic abilities. Models of working memory separate verbal and visuo-spatial processing (Baddeley, 1990). Finally, a great deal of experimental work has addressed both spatial cognition and psycholinguistics, but relatively little bridges between the areas.

A close examination of the extant bridging literature, however, does indicate the usefulness of language in elucidating properties of spatial representations. Indeed, some view spatial knowledge as the basis of linguistic knowledge (Tversky, in press). Others, such as Miller (1990) espouse ideas reminiscent of Coleridge in saying, 'It is significant to note that the device most often used for conversion from private to public is language'. Situations where we use language to describe space are not uncommon; many an individual has been stopped on the street and asked for directions. In reality, the spatial descriptions given in response to direction queries could reflect the underlying representation, the properties of language, or both. The present paper discusses evidence supporting the reflection of spatial representations in spatial language processing.

For language to reveal the nature of a spatial representation, it must include a representational mechanism. Everyone agrees that the primary function of language is communication and, as such, language must include both comprehension and production components. These components need not and indeed cannot be independent; by necessity language production includes a comprehension function, as well as additional mechanisms (Levelt, 1989). During communication, one conceptualizes both ideas to be related and those already related. As such, language comprehension and production models both have a representational mechanism. The 'product' of this mechanism has received extensive focus in language comprehension models and has been referred to using similar, although slightly varied terms, including *mental models* (Johnson-Laird, 1983), *situation models* (Kintsch and van Dijk, 1978), and *knowledge structures* (Gernsbacher, 1990). Representational systems are included, although not discussed to the same extent, in language production models (Fromkin, 1981; Garrett, 1984; Levelt, 1989). The mere fact that language models include representation indicates that language processing may be useful in understanding spatial representation.

What information can be gleaned from spatial description processing? Through language a spatial array can be described in informationally equivalent, yet different ways. Differences arise by invoking different perspectives, for example survey versus route perspective or speaker-centred versus addressee-centred perspective. The nature of these description differences has implications for understanding spatial representations. Properties of the underlying representation affect properties of descriptions based on that representation, although not necessarily in a canonical fashion. Since for a given environment one could take either a route or a survey perspective in a description, what determines the perspective selected? Most likely properties of the underlying representation tip the balance between the two perspectives. Similarly, salient features in the representation should influence both the route selected from amongst a candidate set of routes and the specific landmarks included.

Language can also provide different types and levels of information about a spatial array. A description can include locative and/or non-locative details, both with varying levels of specificity. As an example, locative information can range from broad categorical relations to detailed metric points. A simpler description may contain only categorical locative information, while a more complex one may include both metric locative and additional non-locative details. In contrast, actual environments and their depictions necessarily include multiple types and levels of information. Both metric and categorical information is available, although not necessarily processed. Because the type and level of information can be varied nearly independently in descriptions, questions regarding the necessity and sufficiency of specific information for a mental representation can be addressed. This can be accomplished through both language comprehension and production. For comprehension, descriptions can be constructed to systematically vary the type and level of information; for production, descriptions can be analysed for the type and level of information included.

Spatial descriptions can also be used to relate a variety of different spatial arrays, ranging in complexity from simple object-to-object relations to complex urban environments. Interestingly, array complexity is not directly related to the complexity of issues arising when describing the array. Two objects make up the least complex array. Yet, terms used to describe the relative spatial positions of two objects are often fraught with ambiguity derived either from broad meanings of the terms, such as spatial preposition (Herskovits, 1985), or the existence of multiple reference frames by which the description may be interpreted (Garnham, 1989; Levelt, 1984). With multiple object arrays, other issues arise. Since the relations between multiple objects can take on several different configurations, issues arise over how the relations should be linearized for language. Linearization is necessary as space is generally two- or three-dimensional while the stream of language is uni-dimensional. Thus, individuals devise strategies to translate the multi-dimensional nature of space into language (Levelt, 1982b). Linearization is but one issue for multiple object arrays, but it serves to illustrate the different issues arising in relation to spatial situation complexity. As such, an understanding of how language reflects underlying spatial representations cannot be complete without considering a range of spatial arrays.

Early thoughts on spatial descriptions

The earliest known spatial descriptions reflected properties of the describer's mental representation. Cicero purports that an Ancient Greek poet, Simonides, used a spatial tour to name the victims of a roof collapse, a procedure now referred to as the *method of loci* (Bower, 1970). Although the description itself was most likely spatially sparse, including mainly victims' names, its basis was a gaze tour originating from Simonides' location at the banquet.

More systematic studies of spatial descriptions came in the 1970s. A now classic study by Linde and Labov (1975) had individuals produce connected discourse relating the layout of their New York apartments. This work showed that spatial descriptions bear characteristics easily attributable to underlying spatial representations. More specifically, the resulting descriptions primarily took the form of tours or routes through the apartments. Partially as a result of this work, researchers and theorists assumed that spatial descriptions typically take a route or mental tour format (Levelt, 1982b). This position also takes into account other features potentially affecting spatial representations, particularly how the environment was learned. A mental tour reflects how we commonly experience environments such as apartments – we walk through them (Levelt, 1982a; Levelt, 1982b).

However, we experience other environments differently. One need not tour a single room, as it can be viewed in its entirety from a single vantage point. Ehrich and Koster (1983) collected descriptions of a dollhouse room, finding that they reflected gaze tours rather than actual tours. Gaze tours do not include hypothetical movement through the environment, but instead incorporate movement of the eyes from a fixed point either within or just outside the environment (see also Shanon, 1984). Tours and gaze tours reflect ways we interact with smaller environments. Larger ones can potentially be viewed from a vantage point above the environment. This viewpoint is commonly referred to as a *survey* perspective. Modern technology now affords such views from satellites or planes, but prior to these technologies or even the existence of maps, a well-placed cliff served the same purpose. Because there are different ways to experience an environment, the experience may impact the resultant spatial representations and may in turn become an integral part of an environment description. A predominant use of routes in descriptions may alternatively reflect the best linearization match – routes naturally linearize spatial information.

Like single rooms, object-to-object relations are generally experienced from a single vantage point. Although an object-to-object relation is spatially simple, descriptions of the relationship are not. In the description, 'The box is to the left of the pig', one must know additional information to ascertain the exact locations of the box, including where the speaker (or addressee) is located with respect to the pig and whether the utterance is using the speaker's, addressee's, or pig's viewpoint. In other words, objects must be located with respect to a reference frame, including an origin, a co-ordinate system, terms of reference, and a reference object.

Although the details of and the terminology used for such reference systems have been much debated, a general consensus rests on three possible frames (Garnham, 1989; Levelt, 1984; Levelt, 1989; Levinson, 1996). In Levinson's (1996) terms, the *relative* framework uses a coordinate system centred on one of the conversational participants, speaker or addressee, and relates the location of two external objects from that perspective. The traditional definition of *deictic* fits this framework. The *intrinsic* framework uses a co-ordinate system centred on a particular object and relates a single external object to it. To use this framework, the referent object must have defined intrinsic sides. Finally, in *extrinsic* or *absolute* frames, the co-ordinate system is external to the scene. The most common extrinsic system involves cardinal directions, *north*, *south*, *east*, and *west*, and is more easily distinguished from the frames because it generally uses different terminology. The relative and intrinsic frames both use the terms *left*, *right*, *front*, and *back*, and thus can cause some confusion. In the case of the pig example, the extrinsic frame is not likely to be used, at least in English, leaving the relative and intrinsic frames as competing possibilities (see Levinson, 1996).

In describing object-to-object relations, do speakers default to a particular reference frame? In other words, if one hears, 'The box is to the left of the pig', is a specific box (see Figure 11.1) consistently chosen? In this situation, with the speaker's position corresponding to the view given in the figure, the relative and intrinsic frames conflict. A consistent box selection would signal the default reference frame, which in turn would reflect the most likely spatial representation. There has been little agreement on which reference frame is the default. Some argue for relative (Levelt, 1982b; Levelt, 1989; Linde and Labov, 1975), evoking ease of cognitive

Figure 11.1 Example of object-to-object relations where spatial terms, such as *left* are ambiguous without knowledge of the reference frame used.

computation for support. The relative frame matches the speaker's perception and, as such, requires no additional computation. Alternative frames must be computed, either through mental rotation or other cognitive mechanisms. Others find little contribution of the relative frame (Carlson-Radvansky and Irwin, 1993). Still others argue for the primacy of the intrinsic frame (Miller and Johnson-Laird, 1976), basing their views on salient properties of the objects, such as intrinsic sides. Additionally, an intrinsic frame is easier to use when the object is in motion (Levelt, 1984). Only recently have empirical studies begun to address the question of a default reference frame.

Recent studies of spatial language

The early thoughts and studies on spatial language discussed above afforded fertile grounds for further empirical studies. Discussions of spatial frames left open questions about the existence of a default, particularly because theoretical positions disagreed. Similarly, the format of environment descriptions also opened up questions about default strategies. In this case, however, early thoughts more or less converged on a route format. Default strategies alone, however, do not provide sufficient information to assess the efficacy of using language to infer properties of spatial representations. As a result, other issues in spatial description processing needed to be examined and compared to general findings about spatial representations.

Michel Denis and his colleagues have shown similarities between mental representations developed from spatial analogs and those developed by reading descriptions. Denis and Cocude (1989) showed that mental scanning time for environments learned from descriptions increased as distance between landmarks increased, similar to findings shown by Kosslyn *et al.* (1978) for environments learned from a map. Additionally, individuals show evidence of the symbolic distance effect for environments learned from descriptions, taking longer to estimate shorter distances (Denis and Zimmer, 1992), just as is shown for environments learned from maps (Maki, 1982). These findings suggest that basic properties of spatial representations emerge regardless of the manner in which the spatial information is acquired.

Spatial frames of reference

Recent behavioural studies on spatial frames indicate that no single frame has priority processing in all situations. Instead, results of these studies point to multiple situational influences on the initial interpretation. Aspects of the actual described scene exerted some influence. Carlson-Radvansky and Irwin (1993) found that manipulating the relationship between objects in a scene, such as their physical closeness, influenced frame selection. In their studies, the term *above* was primarily interpreted from an environment-centred perspective, but manipulations calling attention to the objects in the scene led to more object-centred interpretations. In all their cases the environment-centred frame was co-ordinated with the viewer-centred frame.

Friederici and Levelt (1987) showed that uncoupling environment-centred and viewer-centred perspectives via the elimination of gravity resulted in viewer-centred descriptions. Thus, gravity serves as a strong situational cue for environment-centred frame selection. Franklin and Tversky (1990) further supported the gravitational influence. In their studies examining the Spatial Framework Hypothesis, participants read descriptions of a character surrounded by objects. After learning a scenario, the character was described as facing a particular object and response times to access information about the other surrounding objects was assessed. If the individual in the scene was described as standing upright, then response times to objects above and below the individual were fastest. If, however, the individual was described as reclining, then the precedence for objects in the gravitational plane was diminished in favour of objects in front of and behind the character.

Situational variables need not be directly related to the scene to exert an influence; aspects of the broader conversational setting can also affect spatial frame processing. Schober (1993) showed that the presence of a conversational partner influenced spatial frame selection. Participants with a partner used more viewer-centred frames while those with an imaginary partner used more addressee-centred frames. Further, Schober and Bloom (1995) found that speakers initiated utterances faster when speaking from a viewer-centred perspective, next fastest from an object-centred, and surprisingly, slowest from an addressee-centred frame (Schober, 1996). In an even broader

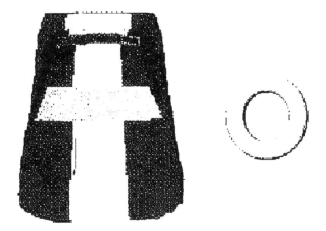

Figure 11.2 Example stimuli from ERP study examining spatial reference frame. Relative and intrinsic frames conflict in example.

context, Levinson (1996) found cultural influences on reference frame selection that were attributable to the availability of spatial terms in a culture's language. Taken together, these findings indicate that the spatial frame processed depends on a wide range of contextual variables and that consideration of both local influences, such as the relative distance between the two objects (Carlson-Radvansky and Irwin, 1993), and global influences, such as culture (Levinson, 1996) should be given.

The evidence for situational influences leaves open the possibility that alternative reference frames receive some level of simultaneous activation and are thus both represented. If so, then the situational variables tip the balance toward the actual reference frame processed. Carlson-Radvansky (Carlson-Radvansky and Jiang, 1998; Carlson-Radvansky and Logan, 1997) and her colleagues have provided some indication for simultaneous activation of multiple reference frames. Using a negative priming technique, Carlson-Radvansky and Jiang (1998) found evidence that both intrinsic and relative reference frames were initially activated. With both frames active, one is then selected for use.

In a series of studies (Faust *et al.*, 1998; Taylor *et al.*, 1998), we have examined spatial frame processing and the possibility of multiple frame activation using psycho-physiological measures, namely evoked response potentials (ERPs). ERPs have a methodological advantage in being able to monitor the time course of cognitive processing. They measure immediate electro-physiological changes in the brain in response to specific stimuli. In our study, participants viewed pictures containing two objects, one with intrinsic sides (car, chair, etc.) and the other without (in this case, a doughnut-shaped figure; see Figure 11.2). The intrinsic object faced one of four directions (front, back, left, or right) and the doughnut lay to one side of it (intrinsically front, back, left, or right). When the object faced the same direction as the participant, the intrinsic and deictic frames aligned; otherwise, they competed.

After viewing the picture, a brief fixation point appeared followed by one of three spatial terms (*left*, *right*, or *front*; the term *back* could not be used as objects could not be displayed behind a participant). Participants then decided whether the term *could* be used to describe the picture, i.e. was it accurate from either a viewer-centred or object-centred frame. Terms were paired with pictures to create four conditions: (1) term was correct for both frames (both correct), (2) term was correct from the intrinsic object's perspective (intrinsic correct); (3) term was correct from viewer-centred perspective (relative correct); and (4) term was incorrect for both frames (both incorrect). Initial activation of both frames implies parallel processing, and would support differential brain wave response for the *both correct* condition compared to the other three conditions. The ERP results gave some indication of initial joint activation of frames. Brain waves corresponding to the both correct condition showed greater positivity between 300 and 450 milliseconds (ms) after presentation of the spatial term (see Figure 11.3). Additionally, the results showed that the intrinsic and deictic frames were processed differently, somewhat tempering the joint activation findings.

Interestingly, the intrinsic frame contributed more to the results than the relative frame. A comparison of conditions manipulating whether the term was intrinsically correct, but holding the relative interpretation constant, found an increased negativity for an incorrect intrinsic interpretation (compare the *correct intrinsic* condition to the *both incorrect* condition in Figure 11.3). This negativity peaked at approximately 325 ms after the spatial term presentation. In contrast, a comparison manipulating the relative interpretation while holding the intrinsic constant found no significant differences to terms that were either correct or incorrect from a relative perspective (compare the *correct relative* condition to the *both incorrect* condition in Figure 11.3). The ERP component we believe is affected by the spatial frame manipulation is the N400, a component associated with semantic integration (Holcomb, 1993). Thus, a term that correctly describes both the intrinsic and relative frame is easily integrated with the information participants processed from the picture. In contrast, a term that was incorrect from an intrinsic perspective appeared to be more difficult to integrate. Although somewhat speculative, this finding implies early joint activation of frames with a follow-up precedence for intrinsic frame processing.

A secondary, although interesting, result indicated that spatial frames are processed quite rapidly. Differences based on our frame manipulation peaked at approximately 325 ms after participants saw the spatial term. While it is possible that participants re-coded the pictorial scene into the

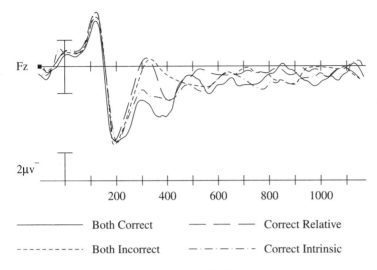

Fz

2μv⁻

200   400   600   800   1000

——————— Both Correct   — — — Correct Relative

----------- Both Incorrect   — · — · — · · Correct Intrinsic

Figure 11.3 ERP results showing spatial frame processing.

appropriate spatial terms for their task, this seems unlikely. To do so they would have to note the appropriate terms for both frames, drawing from the same small pool of terms (*left, right,* and *front*). While both this strategy and an image processing strategy are consistent with our data, comments by experiment participants suggest that they did not verbally re-code the spatial scene. Participants reported maintaining an image of the spatial scene and comparing the term to the image. Thus, processing of the three spatial terms reflects the representation derived from viewing the depiction of the scene.

We consider the spatial situation in our study to be a basic case in which to examine default spatial frame selection. This is not to exclude other types of scenes as 'basic'. The fact that the intrinsic frame drove our findings was somewhat surprising in light of the literature. Although recent empirical findings show situational effects, they also generally support an initial relative interpretation (Friederici and Levelt, 1987; Schober, 1993; Schober and Bloom, 1995). This discrepancy could be a function of our scenes; each contained only two objects, one with intrinsic sides and one without. However, another study addressing this issue (Taylor *et al.*, 1998) used different scenes containing a wider variety of both intrinsic and non-intrinsic objects and found similar results. It seems more likely that it is a function of either our processing measure (ERPs) or our task. Behavioural and electro-physiological measures do not always support the same processing mechanisms (Kounios and Holcomb, 1992). In this case, behavioural measures may better reflect decision mechanisms after the spatial frames have been processed. Additionally, because participants responded to many similar trials that required spatial frame processing, they may have adopted strategies specific to our task. In this case, participants may have focused purposefully on the intrinsic frame while viewing the picture, thereby completing the more cognitively demanding processing while the spatial scene was perceptually available.

Environment descriptions

Environments consist of many object-to-object relations. Just as single object-to-object relations can be described in several ways based on the frame selected, environments can be described in different, yet informationally equivalent ways. As discussed earlier, tours serve as a common way of experiencing an environment, but not the only common way. Additionally tours have the advantage of naturally fitting the linearization requirement of language (Levelt, 1982b). With two factors favouring routes, are they actually the default? Like object-to-object descriptions, environment descriptions (and the underlying representations) could be influenced by situational variables. There is ample evidence supporting situational influences on spatial representations. Factors such as learning by map or navigation (Sholl, 1987; Thorndyke and Hayes-Roth, 1982), the existence of hierarchies (Hirtle and Jonides, 1985; Huttenlocher *et al.*, 1991; Stevens and Coupe, 1978), and the presence of barriers (Kosslyn *et al.*, 1975; Newcombe and Liben, 1982) all impact spatial memory. If spatial descriptions reflect spatial representations, they too should show similar influences of situational variables.

In work with Barbara Tversky (Taylor and Tversky, 1996), we collected descriptions of environments learned from either maps or navigation and examined them for choice of perspective (route or survey). Several factors influenced perspective choice, including learning method (map or navigation) and qualities of the environment (single or multiple paths; similarly or differently scaled landmarks). Thus, situational variables influenced properties of the descriptions, in this case perspective. As such, route descriptions cannot be viewed as an absolute default. Interestingly, in all conditions and for all environments, participants produced route and survey descriptions. Additionally, some descriptions mixed perspectives, either redundantly or for different parts of the environment. This finding is contrary to claims that once selected, perspectives are applied consistently in descriptions (Levelt, 1982a; see also Schober, 1995). Not only are route descriptions not an absolute default, but even if selected, the route perspective may not be used throughout a description.

The fact that participants produced route, survey, and mixed descriptions for all environments also indicates that individuals use spatial representations flexibly. We can produce descriptions in perspectives different from those we have experienced. In our studies (Taylor and Tversky, 1996), participants produced route descriptions after studying maps and survey descriptions after navigating. In real life we follow routes we have seen on maps, or we draw maps of environments we have navigated, such as those provided to attendees of a dinner party. Similar evidence for representational flexibility comes from our earlier studies examining spatial description comprehension (Taylor and Tversky, 1992b). In these studies participants read either a route or a survey description of an environment and then verified inference statements taking either the same or a different perspective. Results showed that participants could process either perspective, making spatial inferences in either perspective with equal speed and accuracy.

In both the above studies (Taylor and Tversky, 1992b, 1996), individuals were to learn the entire environment. What happens when we have a more specific goal for learning spatial information? Studies by Schneider and Taylor (1999) had participants learn routes through an environment, rather than entire environments. All information was related through descriptions taking either a route or a survey perspective. Similar to Taylor and Tversky (1992b), responses to route and survey inference statements were equally fast and accurate, but only if participants read a route description. Unlike the previous work, participants reading a survey perspective description responded faster and more accurately to survey test statements. Survey perspective participants also drew more maps when recalling information about the environment (Study 1) and when taking notes about it for later use (Study 2). The fact that participants' goal was to learn a route may account for this finding. Because the survey perspective is discrepant with the route goal, it may have been more salient. Similarly, Perrig and Kintsch's (1985) route and survey descriptions reflected some properties of routes. Their findings supported maintenance of learned perspective, perhaps because the perspective information was made more salient by the description's route features.

Additional support for goal-directed learning effects come from two studies by Taylor and Naylor (in press). In these studies, participants studied an environment, either by navigating or via a map, with a specific, perspective-based goal. The survey goal had participants learn the overall layout of the environment; the route goal had them learn the fastest routes between locations. A number of memory measures showed that their representations were influenced by both *how* and *why* they learned the environment. For one of these tasks, participants wrote route descriptions. An examination of the description content showed effects of learning goal such that participants who had a route goal included more spatial direction terms. Schneider and Taylor (1999) also found description content differences reflecting the learning goal. In Schneider and Taylor's second study, participants took notes based on an oral route description to use for a later wayfinding task. Participants listened to the description twice, taking notes in different coloured ink each time. An analysis of the note content showed that participants included spatial information relevant to the route goal the first time they listened to the description and then added salient non-spatial details the second time. Thus, both of these studies indicate that the learning goal influenced the underlying representation and how the representation was used to produce descriptions.

In the studies discussed thus far, individuals received equivalent information about an environment. In reality, information can be filtered through language leading to the inclusion of different types and levels of detail. For example, the spatial information on a map could be described in detailed, metric terms (4.13 miles north) or in broad, categorical terms (to the north). While both metric and categorical information are available from maps, both need not be included in a description. Similarly, while spatial information on a map is determined, it can be left indeterminate in descriptions. Mani and Johnson-Laird (1982) had participants read descriptions of simple object arrays that either determined the location of all objects or left some spatial relations indeterminate. For example, if one says, 'The knife is to the right of the plate; the spoon is to the right of the plate', we do not know the relationship between the spoon and the knife. Mani and Johnson-Laird proposed that people form mental models for determinate descriptions, but keep the propositions in mind for indeterminate ones. Their results concur with their proposal, showing that individuals could better identify verbatim lines from indeterminate compared to determinate descriptions (see also Denis and Cocude, 1992).

More complex environments afford more avenues of indeterminacy. As with Mani and Johnson-Laird (1982), landmarks can be underspecified in a description leading to either more than one possible location or absence of pertinent detail. In the former case, the indeterminacy is apparent while processing the description. In the latter, it may not be apparent until one compares their mental representation to the actual environment. For example, if you heard, 'turn when you reach the intersection of Main Street and Tresscott Road', you might assume that turns in only one direction are possible and represent it as such. Only when approaching this intersection and seeing that both left and right turns are possible does the lack of specificity become apparent and important for your representation. For environment descriptions, this latter form of indeterminacy is more likely. Complex environments also fall prey to overdeterminacy in descriptions if too much detail is provided. While undetermined information may not be readily apparent, the sheer amount of spatial information available in an overdetermined description stands out.

The Schneider and Taylor (1999) research also examined effects of different levels of description determinacy on both learning and memory. As discussed earlier, the descriptions took either a route or a survey perspective. In addition, they had one of three determinacy levels (indeterminate, determinate, or overdeterminate). Overdeterminacy affected both learning and memory. Participants reading overdeterminate descriptions spent longer studying, but recalled fewer total ideas and made more errors during wayfinding. Indeterminacy, on the other hand, did not influence learning and influenced memory only when mental representation was compared to the actual environment, such as when using a map for wayfinding. In summary, participants had difficulty developing representations from overdetermined spatial descriptions. Not surprisingly, this difficulty carried over to using the representation. On the other hand, participants had no greater difficulty developing representation from indeterminate descriptions than they did from determinate ones. However, once the indeterminacy was discovered, difficulties arose.

Our findings based on overdeterminacy are not surprising when considering the fact that we employ various heuristics and cognitive economizing strategies when representing complex spatial information. For example, Stevens and Coupe (1978) asked for judgements about relative locations of landmarks, such as 'Which is further west, Reno, NV or San Diego, CA?' In arriving at a response to such a question, we could maintain a representation elaborating pair-wise combinations of locations. Although possible, this solution is unlikely as it requires an extensive amount of information. Alternatively, locations could be associated to larger regions, such as Reno to Nevada and San Diego to California, and a response generated based on the relative locations of Nevada and California. Stevens and Coupe's (1978) results support this strategy. This and other strategies (for a review, see Tversky, in press) keep the information load in check thereby striking a balance between information availability and efficiency of processing. Taylor and Tversky (1992a; 1996) found evidence of these cognitive economizing strategies in spatial descriptions. One study included a complex Amusement Park map with hierarchically organized levels of spatial detail. Descriptions of this environment, written from memory, included an intermediate level of detail grouped into distinct regions. Similarly, participants reading overdeterminate descriptions in the study by Schneider and Taylor (1999) may also have 'simplified' the environment using heuristics.

Taken together, the present research on spatial descriptions, both comprehension and production, illustrates the extent to which qualities of an underlying spatial representation can be illuminated through language. This work covers a range of different spatial arrays and a range of language tasks. While spatial description processing could reflect general language processing, the evidence from spatial description processing points more strongly to aspects of the spatial representation. Addressing this issue more specifically, Taylor and Tversky (1992a) addressed the separate contributions of language processing and underlying representations. We told participants they would *either* write a description relating information from a map *or* redraw the map, both from memory. We used this manipulation to induce expectations about using language, thereby engaging language processing, particularly linearization. Participants actually performed both tasks, writing a description and drawing a map, in counterbalanced order. The manipulation had no effect on linearization, as order of landmarks written in descriptions and placed in depictions positively correlated and did not differ as a function of task expectations.

Future

The research discussed here indicates that spatial language processing reflects aspects of the underlying representation. Thus, the future of research in this area could address unanswered questions about spatial representations. Such broad questions regarding the extent of flexibility in representing spatial information, the manner by which spatial information is integrated and updated, and the level of detail represented remain largely unanswered. Since these broad questions apply to the range of spatial situations discussed thus far, from object-to-object relations to complex environments, their answers would have wide generalizability for understanding spatial cognition. As discussed below, these questions can be approached using spatial descriptions.

Two situations would account for apparent flexibility in representing spatial information, a single representation allowing flexible access or multiple representations simultaneously activated. Both are possible and may depend on the nature of the spatial situation, in particular its complexity. Object-to-object relations may be simple enough to activate more than one representation (Carlson-Radvansky and Irwin, 1994; Carlson-Radvansky and Jiang, 1998; Carlson-Radvansky and Logan, 1997). The ERP evidence on spatial frames (Faust *et al.*, 1998; Taylor *et al.*, 1998) provides some evidence for simultaneous activation frames when viewing a scene. However, the issue is far from resolved. Additional psycho-physiological evidence as well as convergent evidence from comprehension and production paradigms could provide more definitive evidence. Multi-object arrays, such as environments, may be too complex to maintain multiple representations, thus flexible representations may be developed to maximize usability, as suggested by Taylor and Tversky (1992b; 1996).

The degree to which a representation is flexible depends on how spatial information is integrated. However, little is known about this integration process. Whether gleaned from maps, navigation, or descriptions, spatial information about entire environments cannot be processed all at once. To achieve a full, coherent representation, spatial information must be integrated over time. Through maps, information is acquired over successive eye fixations. In navigation and through texts, information is received sequentially. Thus with all input media, integration is necessary. Through text, the order that information about an environment is conveyed can be manipulated, varying the effort needed to integrate individual pieces of information. Current studies are addressing this issue using both maps (Taylor *et al.*, 1999) and descriptions (Taylor and Busch, 1999). By approaching the issue using both analogue and textual input, the results should provide insights into the integration process.

That spatial language processing indicates aspects of underlying representations to a large degree is not to say that the language processing has *no* influence on comprehension and production of descriptions. The extent to which language processing affects spatial representations and the nature of these effects has not been determined. For example, how do representations differ as a function of whether spatial information was learned from a spatial description, from a map, or via navigation? While Taylor and Tversky (1992b) did not find differences in representational perspective between map and description learners, few studies have directly compared analogical inputs to language inputs. Different predictions emerge for pairwise comparisons of the three input modes, largely because these comparisons reveal similarities on different dimensions. Maps and navigation compare with respect to analogical information; descriptions and navigation both provide serial input. Additionally, route descriptions match navigation and survey descriptions match maps when considering spatial perspective. Across these possible comparisons, language input to spatial representations would be apparent only from tasks tapping representational properties unique to descriptions.

One such property involves level of available detail. Descriptions generally contain less detail than maps or actual environments. Detailed spatial information is difficult to coherently relate through descriptions (Passini, 1984; Schneider and Taylor, 1999). Although a greater level of detail is available in maps and through navigation, it is not necessarily processed. Determining the extent to which detail is mentally represented as a function of input medium would indicate one influence of language processing. Such knowledge might also inform us about the correct level of detail to include in a spatial description. Level of detail is only one example of dimensions by which representations based on language and those based on other input media can be compared.

Samuel Taylor Coleridge's statement, 'The best part of human language, properly so called, is derived from reflection on the acts of the mind itself' (Coleridge, 1834) introduced the concept of viewing spatial representations through the lens of language. Theoretical inclusion of representational components in language theories, both comprehension and production, progresses the idea an additional step forward. The evidence from spatial language studies reflecting properties of the underlying representation only serves to improve the focus. Thus, future work using spatial descriptions may further our understanding of how spatial representations are acquired, updated, and used.

References

Baddeley, A. (1990) *Human memory: Theory and practice*. Boston, MA: Allyn and Bacon.

Bower, G.H. (1970) Analysis of a mnemonic device. *American Scientist*, 58, 496–510.

Carlson-Radvansky, L.A. and Irwin, D.E. (1993) Frames of reference in vision and language. *Cognition*, 46, 223–44.

—— and —— (1994) Reference frame activation during spatial term assignment. *Journal of Memory and Language*, 33, 646–71.

—— and Jiang, Y. (1998) Inhibition accompanies reference-frame selection. *Psychological Science*, 9(5), 386–91.

—— and Logan, G.D. (1997) The influence of reference frame selection on spatial template construction. *Journal of Memory and Language*, 37, 411–37.

Coleridge, S.T. (1834) *Biographia literaria*. New York: Leavitt, Lord.

Denis, M. and Cocude, M. (1989) Scanning visual images generated from verbal descriptions. *European Journal of Cognitive Psychology*, 10, 115–43.

—— and —— (1992) Structural properties of visual images constructed from poorly or well-structured verbal descriptions. *Memory and Cognition*, 20, 497–506.

—— and Zimmer, H.D. (1992) Analog properties of cognitive maps constructed from verbal descriptions. *Psychological Research*, 54, 286–98.

Ehrich, V. and Koster, C. (1983) Discourse organization and sentence form: The structure of room descriptions in Dutch. *Discourse Processes*, 6, 169–95.

Faust, R.R., Naylor, S.J., Taylor, H.A. and Holcomb, P.J. (1998) *Distinct processing of spatial reference frames: Electrophysiological evidence*. Paper presented at the 5th Annual Meeting of the Cognitive Neuroscience Society, San Francisco, CA.

Franklin, N. and Tversky, B. (1990) Searching imagined environments. *Journal of Experimental Psychology: General*, 119, 63–76.

Friederici, A.D. and Levelt, W.J.M. (1987) Resolving perceptual conflicts: The cognitive mechanisms of spatial orientation. *Aviation, Space, and Environmental Medicine*, 58, 164–9.

Fromkin, V. (1981) The nonanomalous nature of anomalous utterances. *Language*, 47, 27–52.

Garnham, A. (1989) A unified theory of the meaning of some spatial relation terms. *Cognition*, 31, 45–60.

Garrett, M.F. (1984) The organization of processing structure for language production. In Caplan, D., Lecours, A.R. and Smith, A. (eds), *Biological Perspectives on Language*. Cambridge, MA: MIT, pp. 172–93.

Gernsbacher, M.A. (1990) *Language Comprehension as Structure Building*. Hillsdale, NJ: Erlbaum.

Herskovits, A. (1985) Semantics and pragmatics of locative expressions. *Cognitive Science*, 9(3), 341–78.

Hirtle, S.C. and Jonides, J. (1985) Evidence of hierarchies in cognitive maps. *Memory and Cognition*, 13, 208–17.

Holcomb, P.J. (1993) Semantic priming and stimulus degradation: Implications for the role of the N400 in language processing. *Psychophysiology*, 30, 47–61.

Huttenlocher, J., Hedges, L.V. and Duncan, S. (1991) Categories and particulars: Prototype effects in estimating spatial location. *Psychological Review*, 93(3), 352–76.

Johnson-Laird, P.N. (1983) *Mental Models*. Cambridge, MA: Harvard.

Kintsch, W. and van Dijk, T.A. (1978) Toward a model of text comprehension and production. *Psychological Review*, 95, 163–82.

Kosslyn, S.M., Ball, T.M. and Reiser, B.J. (1978) Visual images preserve metric spatial information: Evidence from studies of image scanning. *Journal of Experimental Psychology: Human Perception and Performance*, 4, 47–60.

—— Pick, H.L. and Fariello, G.R. (1975) Cognitive maps in children and men, *Child Development*, 45, 707–16.

Kounios, J. and Holcomb, P.J. (1992) Structure and process in semantic memory: Evidence from event-related brain potentials and reaction times. *Journal of Experimental Psychology: General*, 121, 459–79.

Levelt, W.J.M. (1982a) Cognitive styles in use of spatial direction terms. In Jarvella, R.J. and Klein, W. (eds), *Speech, Place, and Action*. New York, NY: Wiley, pp. 251–68.

—— (1982b) Linearization in describing spatial networks. In Peters, S. and Saarinen, E. (eds), *Processes, Beliefs and Questions*. Dordrecht: Reidel, pp. 199–220.

—— (1984) Some perceptual limitations on talking about space. In van Doorn, A.J., van de Grind, W.A. and Koenderink, J.J. (eds), *Limits in Perception*. Utrecht: VNU Science, pp. 323–58.

—— (1989) *Speaking: From Intention to Articulation*. Cambridge, MA: MIT.

Levinson, S.C. (1996) Frames of reference and Molyneux's question: Crosslinguistic evidence. In Bloom, P., Peterson, M.A., Nadel, L. and Garrett, M.F. (eds), *Language and Space*. Cambridge, MA: MIT, pp. 109–70.

Linde, C. and Labov, W. (1975) Spatial networks as a site for the study of language and thought. *Language*, 51, 924–40.

Maki, R.H. (1982) Why do categorization effects occur in comparative judgment tasks? *Memory and Cognition*, 10(3), 252–64.

Mani, K. and Johnson-Laird, P.N. (1982) The mental representation of spatial descriptions. *Memory and Cognition*, 10, 181–7.

Miller, G.A. (1990) The place of language in a scientific psychology. *Psychological Science*, 1, 7–14.

—— and Johnson-Laird, P.N. (1976) *Language and Perception*. Cambridge, MA: Harvard.

Newcombe, N. and Liben, L. (1982) Barrier effects in the cognitive maps of children and adults. *Journal of Experimental Child Psychology*, 34, 46–58.

Passini, R. (1984) Spatial representations, a wayfinding perspective. *Journal of Environmental Psychology*, 4, 153–64.

Perrig, W. and Kintsch, W. (1985) Propositional and situational representations of text. *Journal of Memory and Language*, 24, 503–18.

Schneider, L.F. and Taylor, H.A. (1999). How do you get there from here? Mental representations of route descriptions. *Applied Cognitive Psychology* 13, 415–41.

Schober, M.F. (1993) Spatial perspective-taking in conversation. *Cognition*, 47, 1–24.

—— (1995) Speakers, addressees, and frames of reference: Whose effort is minimized in conversations about locations? *Discourse Processes*, 20, 219–47.

—— (1996) Addressee- and object-centered frames of reference in spatial descriptions. In Olivier, P.L. (ed.), *Cognitive and Computational Models of Spatial Representation: Papers from the 1996 AAAI Spring Symposium*. Menlo Park, CA: AAAI.

—— and Bloom, J.E. (1995) The relative ease of producing egocentric, addressee-centered, and object-centered spatial descriptions. Paper presented at the Annual Meeting of the Psychonomics Society, Los Angeles, CA.

Shanon, B. (1984) Room descriptions. *Discourse Processes*, 7, 225–55.

Sholl, M.J. (1987) Cognitive maps as orienting schemata. *Journal of Experimental Psychology: Learning, Memory, and Cognition*, 13, 615–28.

Stevens, A. and Coupe, P. (1978) Distortions in judged spatial relations. *Cognitive Psychology*, 10, 422–37.

Taylor, H.A. and Busch, C. (1999) Processing continuous and discontinuous descriptions of environments. Unpublished data.

—— and Naylor, S.J. (in press) Goal-directed effects on processing a spatial environment: Indications from memory and language. In Olivier, P. (ed.), *Spatial Language: Cognitive and Computational Aspects*. Dordrecht: Kluwer.

—— , —— and Holcomb, P.J. (1998) Different aspects of scene descriptions lead to differential processing. Paper presented at the 9th Annual Meeting of the Winter Text and Cognition Conference, Jackson Hole, WY.

—— Soraci, S.A. and Cruess, L. (1999) Putting the pieces together: Generative integration of spatial information. Unpublished data.

—— and Tversky, B. (1992a) Descriptions and depictions of environments. *Memory and Cognition*, 20, 483–96.

—— and —— (1992b) Spatial mental models derived from survey and route descriptions. *Journal of Memory and Language*, 31, 261–92.

—— and —— (1996) Perspective in spatial descriptions. *Journal of Memory and Language*, 35, 371–91.

Thorndyke, P.W. and Hayes-Roth, B. (1982) Differences in spatial knowledge acquired from maps and navigation. *Cognitive Psychology*, 14, 560–89.

Tversky, B. (in press) Remembering spaces. In Tulving, E. and Craik, F.I.M. (eds), *Handbook of Memory*. New York: Oxford University Press.

12 Sex, gender, and cognitive mapping

Carole M. Self and Reginald G. Golledge

If we knew what it was we were doing it would not be called research, would it?

Albert Einstein

Introduction

A great deal of controversy about the existence, origin, and importance of sex- and gender-related differences is present in cognitive-mapping and other scholarly literatures. This chapter explores the interconnections between sex, gender, and cognitive-mapping ability from a number of complementary disciplinary perspectives. Only research findings directly relevant to cognitive mapping skills will be reviewed, and recent studies, in particular, will be emphasized.

Throughout the chapter it must be remembered that although our primary focus is on gender-related similarities and differences, gender never exists as a social category alone. Of necessity, each of us must consider the myriad ways in which gender interconnects with other significant social categories such as culture, race, ethnicity, age, sexuality, socio-economic status, educational level, and ability or disability, to affect the knowledge developed from cognitive mapping and the application of spatial abilities.

Do we mean 'sex' or should it be 'gender'?

What are the intended meanings of the words 'sex' and 'gender'? Scholars disagree both about how best to define these commonplace terms and about the theoretical bases and research methodologies most appropriate to be used for exploring and evaluating sex- or gender-related effects. According to Crawford *et al.* (1995) 'sex is defined as biological differences in genetic composition and reproductive anatomy and function. Gender is what culture makes out of the "raw material" of biological sex' (p. 341). Gender is generally viewed as being socially constructed and thus represents the 'attributions and meanings a society gives to the concepts of woman and man, girl and boy' (Lott, 1997, p. 280). Researchers in a number of disciplines have asserted that many genders are possible – not just two. They argue that being a woman and being a man are different from generation to generation, and differ for specific racial, ethnic, social, sexual orientation, and religious groups. One thing is for certain: no matter which definition each of us derives to represent our own best view of what the words 'sex' and 'gender' mean, we are still stuck using the terms various authors chose as appropriate when first publishing their research results. The attention recently given to clarification of key terminology should result in a more careful and consistent use of the words 'sex' and 'gender' – if not in everyday discourse, at least when reporting research findings.

Review of past studies

Evolutionary theories and data

Using logic and deduction, a growing number of studies have looked to the evolutionary past for insights into adaptations that would have been favoured by natural selection. For both sexes and from the earliest times, being spatially aware and able to act with reflex speed often meant the difference between life and death (Hall, 1966).

Sex differences in spatial abilities are said to have evolved as a function of the historical division of labour (Tooby and Cosmides, 1992; Tooby and DeVore 1987; McBurney, 1997), with males in most cultures assuming primary responsibility for hunting activity, and females for food gathering. Males' home ranges were found to be larger than females', ostensibly because of their experience with habitat navigation during hunting, warfare, and at least in some cultures, their search for multiple mates. As a result of their hunting and warfare activity, males were said to have developed an implicit understanding of the geometric relationships among objects in physical space (Geary, 1996). They developed the ability to integrate novel incoming spatial information quickly (Geary, 1995), they were well skilled at constructing projectiles, and they exhibited proficiency at aiming at a target (Fischer *et al.*, 1994; Chagnon, 1977). Females, on the other hand, developed spatial skills primarily to forage for food. These skills involved recognition and recall of landmarks and their spatial configurations, peripheral perception, and incidental memory for objects and their locations (Eals and Silverman, 1994). Historically, females have been characterized as being better able than males to record details of their environment through an unconscious imaging process necessitated by their evolutionary roles as caretakers of the young and food gatherers. Buss (1995) postulates that modern humans are descended from women who solved the adaptive challenge of securing a reliable supply of food to carry them through pregnancy and lactation. The evolutionary solution was to prefer mates who demonstrated high cognitive mapping ability to secure resources and a willingness to share them. Geary (1995) indicates that in an evolutionary scale 'sex differences in children's play are viewed as providing the foundation for later cognitive sex differences, based on the position that sex differences in children's play reflect more fundamental differences in the goal structures of males and females' (p. 299).

Cross-cultural theories and studies

Numerous studies have searched for evidence of cross-cultural consistency to provide support for the biological basis (sex hormones, genetics, brain structure and function – including lateralization and handedness) of sex differences in spatial ability (see Berenbaum *et al.*, 1995; Kimura and Hampson, 1994; Grimshaw *et al.*, 1995; Hines *et al.*, 1992; Annett, 1985 and 1994; and McGee, 1982, among others).

Silverman *et al.* (1996) gave the Vandenberg-Kuse test of mental rotations and a space relations test to university students in Canada and to ninth graders in Japan. Sex differences were found to be similar between cultures (large sex effects for mental rotations and small to medium sex effects for spatial relations), indicating to the authors' that hormonal theories of difference may be in play.

Born *et al.* (1987), who conducted a meta-analysis to look at sex, culture, and cognitive abilities among Western, African, and Asian peoples, found strong cross-cultural similarities with respect to sex-related differences in spatial abilities. The study was somewhat constrained in its value, however, because it included no tasks that measured spatial perception and too few three-dimensional tests of mental rotation.

Amponsah and Krekling (1997) undertook research to examine similarities and differences in patterns of sex differences on tests of visual–spatial ability in Ghana and Norway. Two mental rotation tasks (PMA Space test and Vandenberg-Kuse test), and one test each of spatial perception (Water Level test) and spatial visualization (Surface Development test) made up the test battery to be given. Males were found to out-perform females on three (Water Level, PMA Space, and Vandenberg-Kuse) of the four visual–spatial ability tests. The test of spatial visualization showed no sex difference.

A three-dimensional spatial ability Cube Comparison test (3DC, Gittler, 1990) was administered to 384 US and 307 Austrian university students by Tanzer *et al.* (1995). US students were found to work faster but score considerably lower than Austrian students. Two possible reasons for this finding were mentioned: that American students had motivational factors associated with non-serious test-taking behaviour, and the US students did not gain as much practice from working a sample problem as did the Austrian students. Other interesting findings for both countries were that males, participants with technical backgrounds, participants who worked longer on the test items, and participants with higher reasoning and/or higher concentration levels scored higher on the Three-dimensional Cube test.

Socio-cultural influences

Theories based on educational differences

Huang (1993) selected the People's Republic of China as a site to investigate gender differences in cognitive abilities (memory, spatial, reasoning, and verbal factors) among Chinese high school students for the following reasons: 'women are expected to work and share equally in family support which results in more equal career expectations and educational goals for both men and women; the national policy of one child per family has the potential for more equal parental encouragement for both boys and girls; and one educational curriculum is provided for all school children, which ensures each student takes the same mathematics and science courses' (p. 719). Girls (16–17 years) significantly out-performed boys on a Word Span task indicating superior memory ability, on the Word Knowledge test, and on the Computational Speed and Accuracy test (verbal composites); boys out-performed girls only on the Paper Folding test; no gender differences were found on Mathematical Thinking and other reasoning tests. Four of the 11 tests found gender differences, and the effect sizes were not large. Despite boys' better spatial task performance, the author concludes that results seem to support the hypothesis that where social conditions are more uniform, gender differences on visual-spatial and mathematical reasoning skills would be small.

Gender differences in spatial abilities and spatial activity among university students in an egalitarian educational system was the subject of a study by Nordvik and Amponsah (1998). Study participants were 161 Norwegian technology students, and 293 social science students. Four spatial ability tests were administered: Thurstone and Thurstone's (1947) Spatial Relations test, the Vandenberg Mental Rotation test (1978), a Surface Development test, and the Water Level test. Additionally, a thirty-eight-item spatial activity questionnaire was constructed which ended up being similar in item content to earlier activity questionnaires (i.e., Newcombe *et al.*, 1983; Signorella *et al.*, 1986; and Self *et al.*, 1995, and in progress). Nordvik and Amponsah found no support for the claim that gender differences in spatial abilities are declining in modern societies. The study's findings were that male students scored higher in all the spatial ability categories than did females. This was true for both student groups. Male technology students were found to score highest, whereas female social science students scored lowest on the tests. Results of the spatial activity questionnaire revealed gender-typed spatial activity patterns, leading the authors to assert that 'gender typing may be more acceptable in leisure activities than in education and work' (p. 1023).

Two experiments were undertaken by Richardson (1994) to test the effect of different educational levels on a male-typed and a disguised female-typed version of the same mental rotation task. In the first experiment, male undergraduate students performed slightly better than female students; no gender difference was found in the performance of advanced undergraduates and postgraduates; and no performance differences were found in comparing performance between the two versions of the test. Richardson then administered a same–different judgement mental rotation task. On this task the largest gender difference was produced by a postgraduate student sample.

Beliefs about self and about gender groups

Beliefs about self and about gender groups, and the effect these beliefs might have on the spatial performance of women were investigated by James and Greenberg (1997). There are many examples from research studies that spatial ability is commonly viewed as a masculine attribute by both women and men (Meehan and Overton, 1986; Harris, 1981), and that high self-esteem is also associated with masculinity in men and women (James and Greenberg, 1997; Whitley, 1983). In James and Greenberg's study (1997), college undergraduates rated 'remembering the position of several objects relative to each other' as being a masculine task. Building from a rich social-psychological literature on stereotyping (e.g., Deaux and Major, 1987; Steele, 1997), and tying self-esteem both to gender-role orientation and to task performance, the authors focus on social category memberships linked to expectations about particular abilities. Two versions of the task were given. One stressed comparisons between 'men and women', whereas the other indicated comparison to performance 'of other individuals'. The authors hypothesised that because spatial ability is perceived as being stereotypically masculine, and if self-focus does indeed promote seeing self as prototypical of an in-group (James and Cropanzano, 1990), self-focus should lead low self-esteem participants to see the between sex comparison as likely to lead to an outcome unfavourable to women. The study's results provided evidence that 'self-stereotyping in line with gender stereotypes can occur and that it can interact with task type and perceived gender GC [comparison condition] to influence performance' (p. 422).

Brosnan (1998) undertook a study to 'evaluate the effect of describing a spatial ability test either as a measure of spatial ability (traditional format) or as a measure of empathy, upon male and female performance' (p. 203). Participants were eighty-four college students. The Group Embedded Figures Test (GEFT, Witkin, *et al.*, 1971) was given. Half the participants received test instructions emphasizing the spatial nature of the task; the others received test instructions stating the test was a measure of empathy. The Bem Sex Role Inventory (BSRI; Bem, 1981) was also given. Findings were that masculine sex-types outperformed feminine sex-types irrespective of biological sex. Males were not affected by the two different test conditions and performed equally well on each; females, on the other hand, experienced enhanced performance equal with males when testing was done using the 'empathy' rather than 'spatial' version of the instructions. The author states that 'had the [GEFT] measure been distributed in its traditional format the results would have reflected another instance of male superiority in spatial ability traditionally reported in the literature . . . Had only the empathy condition been investigated, no significant sex differences would have been identified' (Brosnan, 1998, p. 211).

Knowledge gained from maps versus that gained from real world experience

Two experiments were undertaken by Roskos-Ewoldsen *et al.* (1998) to explore differences between spatial knowledge gained from maps and that gained from real world navigation experience. The investigators had some doubt about the common belief that mental representations of small spatial layouts (maps) are orientation dependent, and that mental representations of large-scale spaces may be orientation independent (Presson *et al.*, 1989; Presson and Hazelrigg, 1984). The experiment was designed to explore the spatial memory alignment effect for large and small layouts (four-point paths) in a locative condition and non-locative condition. Women were found to be less accurate on judgements of relative direction, indicating that they had a poorer sense of direction than the men who participated in this experiment.

Wayfinding

A number of studies have explored gender-related differences in wayfinding skills. On the side of no significant gender differences are studies on accuracy of pointing to landmarks (Montello and Pick, 1993; Sadalla and Montello, 1989), on wayfinding actions of visitors to an office building (Beaumont *et al.*, 1984), and on wayfinding by the blind (Passini *et al.*, 1990). Men have been found to be more accurate than women in geometric placement of buildings on a map (McGuiness and Sparks, 1983), in locating the direction of landmarks (Bryant, 1982; Galea and Kimura, 1993; Holding and Holding, 1989), in estimating travel distances (Holding and Holding, 1989), and in using cardinal reference points to give directions (Ward *et al.*, 1986). Men made fewer errors in a computer simulation wayfinding task (Devlin and Bernstein, 1995) and men made fewer errors and required fewer trials to learn a novel route than did females (Galea and Kimura, 1993). A positive correlation between route learning and mental rotation task performance was found by Galea and Kimura (1993). Women have been said to be more likely than men to refer to landmarks when giving directions (Miller and Santoni, 1986); they are also more accurate in the recall of landmarks (Galea and Kimura, 1993), and association of objects with a particular location (McBurney, 1997; Silverman and Eals, 1992). Lawton (1994) found that men more frequently say they keep track of their position relative to distant reference points (implying configurational knowledge), whereas women express a preference for step-by-step instructions about a route they are to follow. Women have been characterized as favouring a landmark oriented approach to wayfinding; men are said to favour a spatially oriented approach. Sketch map drawing accuracy has been found to be a good predictor of wayfinding performance (Rovine and Weisman, 1989).

The question 'are there gender differences in indoor wayfinding' was the subject of a 1996 study by Lawton *et al*. The study, using a mixed age group of participants, was designed to measure incidental learning about an unfamiliar environment (university lecture hall). Paths used were coded into one of four patterns: shortcut, partial retrace, exact retrace, and random. Participants electing exact retrace and random strategies made significantly more landmark comments than the partial retrace group. Participants in the exact retrace group showed the highest level of directional accuracy, whereas the random group frequently backtracked. The researchers found that 'there was no significant gender difference in wayfinding pattern, nor was there a difference in time to complete the wayfinding task' (Lawton *et al*., 1996: 214). Men were found to be significantly more accurate in indicating the direction of the destination. Interestingly, women in their self comments reported a marginally greater intention to explore than did men. Lawton *et al*. (1996) link women's greater uncertainty in this study to stereotyping of environmental wayfinding as masculine (Harris, 1981) and the tendency of women to underestimate their performance on tasks that were cross-gender-typed (Beyer, 1990).

Girls and boys, aged ten to seventeen years, participated in a wayfinding behaviour study in a complex walk-through maze (Schmitz, 1997). The researcher was interested in gender-related environmental strategies and the effects that task anxiety and task-specific fear might have on children's spatial performance. Findings were as follows: boys moved through the maze more quickly than did girls; participants who scored higher on the anxiety and fear levels walked through the maze more slowly; girls described themselves as being more anxious than boys on the anxiety and fear measures; no sex differences were found in the total number of elements that were recalled in the maze representation; gender differences were found in strategy preferences, with boys preferring directions in maps and descriptions and girls preferring more landmarks, especially in written descriptions. The author notes this important proviso related to spatial anxiety and choice of environmental strategy: 'It has to be stressed that there is a great variability within the sex groups and that individual differences often superimpose sex differences' (Schmitz, 1997: 225).

A large study (N = 978) was recently undertaken by Malinowski and Gillespie (1998) to examine differences in performance on a real world, large-scale wayfinding task. The men and women in the study had equal motivation to perform well as successful performance was necessary to graduate from the US Military Academy and be commissioned as an army officer. The mandatory military land navigation training took place in a mixed forest area with some clearings. Each participant had to successfully navigate a four hour, ten point for score (about 6,000 metres) course. Prior to being tested, participants received fourteen hours of instruction and practice on using a basic map and compass, as well as instruction on terrain visualization and distance estimation. The second day of preparatory training consisted of buddy-team and individual practice on a 2,000–3,000 metre course. The test day began with a two hour, five point practice course (approximately 3,000 metres), followed by the final four hour test. Findings were as follows: men were found to perform significantly better on the task than women; women took significantly more time to complete the course than did men; men and women who finished the task in the same time range also had significant differences in their task success; self-reported degree of nervousness and map skill ability predicted performance on the task; women reported significantly higher levels of nervousness than did men; previous experience with relevant environmental activities (e.g., orienteering, camping, backpacking, etc.) differed between men and women, with women having significantly less experience with the eight relevant activities; and math scores in class and on standardized tests appear to be correlated with spatial performance. An interesting cultural difference was also found. Participants from bilingual homes found fewer correct points on the task and had slower task completion times. Both differences were significant. When comparing the effects of coming from a bilingual home by gender, women from bilingual homes fared much worse on task performance than did men. Bilingual women found significantly fewer points than women from monolingual homes. The authors note that 'many of the predictors discussed in the spatial ability literature for laboratory or meso-scale spatial tasks are present at this larger-scale. Although it is difficult to isolate the particular spatial skills that may account for the differences, the fact remains that the same differences are being noticed in a complex, real-world spatial task as in laboratory paper-and-pencil tests' (Malinowski and Gillespie, 1998: 16).

Direction giving

Whether there are gender differences in direction giving was a question pursued by Brown et al. (1998). Differences in the use of landmark, relational turn, and cardinal direction cues were measured in a map-present condition to evaluate whether gender differences in females' use of landmarks for navigating (Miller and Santoni, 1986), and males' employment of cardinal directions to navigate more often than females (Ward et al., 1986), might hold for a mixed age population. Two maps (urban and rural) were used; a map was placed in front of each participant, she/he was told the points of origin and destination and instructed to provide directions to a hypothetical stranger to the named destination point. No time restrictions were placed on participants to study the map, and they were told that there were no right or wrong routes to the final destination point. Directions given were scored as being one of four strategy types: landmark, road-name, relational (directional indicators), and cardinal. No significant main effects for gender were found in the amount of study time for the urban or rural maps. Women and men did not differ in their overall use of strategies for direction giving; it was only on the urban map that a gender-related difference for frequency of strategy use was found. Middle-aged females gave the most directions relative to any other age group, and they also studied the urban map longer than any other group. In terms of strategy accuracy, females were less accurate than males only when employing the relational strategy. A reason for this finding may have been that the direction giver was required to make a reversal in the relational turns depending on the route chosen. The authors concluded that 'the absence of errors for the land-mark strategy suggests that this strategy may be more beneficial for navigation when compared to other strategies' (Brown *et al.*, 1998: 141). A call for additional research on the use of landmarks as a navigation strategy is suggested.

Sketch mapping

O'Laughlin and Brubaker (1998) used 160 college students in a self-report task of spatial abilities as well as performance on two spatial tasks – the Standard Mental Rotations test and a cognitive mapping test, to examine if gender differences favouring men existed. Using the floor plan of a home as the task environment, they asked participants to view a brief videotape of the interior of the home and to draw a sketch map of its floor plan. Results revealed a gender difference favouring men for the mental rotation tasks, but no gender differences were found in the cognitive mapping task. The researchers then used a 'landmarks present' condition in which furniture was scattered throughout the house, and a 'no landmarks' condition in which the house was bare, and found no difference between men and women in their performance on a cognitive mapping task. Women, however, reported using a route strategy more often than did men, and they expressed less confidence in the accuracy of their drawn floor plans.

Advanced technology and computer-generated environments

Several studies have investigated the technological innovations that have made navigation in computer-generated 'virtual' environments possible. The escalation of multiple varieties of advanced technology can serve to challenge humans' production and use of cognitive maps. Computer-related studies require competence with technology. Use of technology has been perceived as a masculine activity (Colley *et al.*, 1994; Piller, 1998), video games have been found to be less popular with females than males (Braun and Giroux, 1989; Moffat *et al.*, 1998), and computer anxiety is more frequently found for females than males (Brosnan and Davidson, 1994).

Possible sex differences in spatial route learning in a maze were investigated by Moffat *et al.* (1998). Correlations between maze performance and psychometric tests of spatial ability (Vandenberg Mental Rotations test (1978), Money Road Map test (1965), Guilford-Zimmerman Spatial Orientation test (1956), and two verbal ability tests) were examined. Because of possible effects of computer game experience on spatial performance (Okagaki and Frensch 1994; Subrahmanyam and Greenfield, 1994), participants were asked to rate their experience playing computer games that involve navigating through simulated three-dimensional environments. Findings were that males solved the mazes significantly faster than females across all trials, males made fewer errors than females, and males reported significantly more computer/video game experience than females (although the male advantage was not attributed to greater computer game experience since the use of this variable as a covariate did not eliminate the sex difference on the task). All participants were found to be learning the route through the mazes with repeated exposure; males and females did not differ in the number of information errors on trial 1. On the psychometric tests, no significant gender differences were found on the vocabulary tests; males out-performed females significantly on the Vandenberg Mental Rotations test, the Guilford-Zimmerman Spatial Orientation test, and in the time taken to complete the Money Road Map test. Correlations between maze scores and cognitive test scores showed different patterns for males and females. For females, both spatial and verbal factor scores were significantly correlated with maze performance for completion time and maze errors, leading the authors to question whether females may have been utilizing some verbal strategies to solve the maze. Spatial factor scores for males were correlated with maze completion time and maze errors. Verbal factor scores were not significantly correlated with males' maze performance.

Postma *et al.* (1998) examined sex differences in object location memory. Twenty males and twenty females were used in a task that involved viewing a square with ten different objects on it on a PC screen. After thirty seconds, the objects disappeared from the screen and reappeared in a row above the square. The task was to relocate them in three conditions: an object to position assignment condition in which the original locations were pre-marked on a square; a positions-only condition in which all objects had the same identity and recall of location for them was the primary goal; and a combined condition in which all objects were to be replaced in the square at their original locations without any pre-marked positions being shown. The absolute displacement between original and estimated position was taken to be an indication of success of the integration mechanism used by subjects. Females did as well as males in the object to position assignment condition and on the absolute displacement in the combined condition. They were less efficient than males in positional reconstruction (i.e. where all the objects were the same and only original positions were recalled). The authors argued, therefore, that the observed male advantage was related only to specific processing components and that when encoding involved dual tasks, no significant sex-based differences were observed.

McGivern *et al.* (1998) used a selection of abstract shapes and nameable objects to determine whether gender differences existed in the incidental learning and visual recognition memory of males and females. Using 246 undergraduates, they found that females recognized significantly more abstract shapes and nameable objects than males. In particular, they had subjects locate a star that appeared at various locations on a computer screen amid a background of common objects. Again, females identified significantly more of these incidentally experienced background objects than did males, indicating that even in an incidental environmental learning situation, females were more aware of objects than were males. This ties in with other results that have shown that females observe more details of objects in their local environments, and that their cognitive maps and externally represented spatial layouts are more likely to be richer in detail than those of males.

The growth in route guidance systems in automobiles prompted a study to explore how route guidance information will affect cognitive maps (Jackson, 1996). Learning new routes and being exposed to new geographic territory should enrich and expand one's cognitive map. The Jackson study, in contrast to some previous research (e.g., van Winsum *et al.*, 1990), found significant differences between age, gender, and driving ability groups indicating differences in their ability to cope with and/or use information provided by route guidance and navigation systems.

Geographic knowledge

Students from ages thirteen to twenty-two were the subject of a large study (1,564 students) on gender-related knowledge variations within geography (Henrie *et al.*, 1997). Students were tested about their knowledge of four sub-fields of geography – physical, human, regional, and map skills. A sixty-item test was developed by the authors consisting of questions of an intermediate level of difficulty. Five student groups were established: junior high, senior high, college education/psychology, college introductory geography, and upper-division geography. Gender differences were found with regard to geographic knowledge. Findings were that 'males outperformed females in all of the five student groups and within each of the four subfields, averaging 13 per cent higher' (Henrie *et al.*, 1997: 611). Gender differences in knowledge (14 per cent) were most pronounced in the introductory and advanced geography groups. Evidence was not found that variability in number of geography classes taken, hours spent watching geography-oriented television programming, and current events contributed to geographic knowledge differences. The best predictions of geographic knowledge from this study were found to be hours of reading, number of sisters, ACT scores, and gender. The authors make two interesting observations: 'These findings argue against explaining the gender gap as symptomatic of differences in a single underlying skill such as spatial reasoning . . . It should be stressed that this study did not indicate that males outperform females in classroom situations . . . In fact, a comparison of final grades from a sample of geography classes . . . showed no significant difference between males and females in classroom performance' (Henrie *et al.*, 1997: 618).

Another recent study (Dabbs *et al.*, 1998) found gender differences favouring men in geographic knowledge (world geography). The Dabbs *et al.* study also looked at spatial ability and navigation strategy among college-aged men and women (n = 194). The Vandenberg and Kuse test of mental rotation was given; an object location memory test of forty-eight common items similar to that used by Silverman and Eals (1992); four maps similar to those used by Choi and Silverman (1997) were used to measure local navigation strategy; and world map knowledge was measured with a page-sized world map with fourteen countries to place. Findings were reported as follows: women and men did not differ in object location memory; men excelled at mental rotation; men knew more than women did about world geography; men and women differed in their navigation strategies with men using miles and cardinal directions and women using landmarks and left-right directions; and navigation strategy was not related to world geography knowledge. The authors state that their findings are consistent with Silverman and Eals' (1992) study favouring an evolutionary basis of sex-specific development of spatial skills.

Two case studies related to completion of cognitive mapping tasks were undertaken by Kitchin (1996). In Study 1, college undergraduates from first-year geography practical classes were recruited to investigate their knowledge of a city in Wales and its surrounding area. Thirteen tests were used representing four different information groups: graphic tests involving sketch mapping, partially graphic tests (spatial cued response and cloze procedure tests), uni- to multi-dimensional tests (multi-dimensional scaling and projective convergence), and a test of environmental recognition. In Study 2, one group undertook sketch map 2 and cloze procedure test 2, while other participants undertook the projective convergence test and the orientation specification test. All tests were completed using a think aloud procedure (Newell and Simon, 1972).

Results from both studies revealed very few differences between females and males in geographic knowledge or spatial ability, leading the author to suggest that 'by the age of 18, given no age, education and social differences, males and females have equal cognitive mapping knowledge and ability' (Kitchin, 1996: 285). Kitchin's results agree with results previously found by Golledge *et al.* (1993, 1995) in studies involving female and male geographers and non-geographers performing route learning, distance estimation, and angle estimation tasks.

Cognitive mapping research today

To better understand, measure, and predict women's and men's spatial interests, competence, and cognitive mapping abilities, Golledge, Montello, and Self (1993) began a multi-year, multi-phase study utilizing participants from a two per cent random sample of California voters in the Santa Barbara area.

Unpaid respondents (257 females, 165 males) completed a two-hour, eight-page survey to collect extensive information on each participant's family background, educational experience, wayfinding experience, recreational habits, career expectations, socio-economic status, and so on, from childhood through adulthood. Next, a task involving the sex-typing of spatial activities based on the prior work of Newcombe *et al.* (1983) and Signorella *et al.* (1986), was administered. This was followed by an intensive study of a sub-group of forty-three females and thirty-six males selected from the previously examined larger population. This sub-group performed a large battery of spatial and geographic tasks and underwent psychometric testing of spatial abilities.

We examine two groups of results from this effort. First, we examine differences found in the sex-typing of activities having a spatial component. Next, we report the results of the psychometric, spatial, and geographic testing.

Sex-typing of activities with a spatial component

In a study by Newcombe *et al.* (1983), eighty-one sport and free-time activities having a spatial component were classified by participants as either traditionally masculine (M) or feminine (F). In our replication study (Self *et al.*, 1995; and Self *et al.*, in progress), we hypothesized that sex-typing will have declined because of the last decade's emphasis on equal opportunity and unisex activities, that older participants will sex-type more activities than will younger participants, and that male respondents will sex-type more activities overall than will female respondents.

Although no direct comparisons can be made with the Newcombe *et al.* study because we added 'neutral' as a third classification option, a different social consensus about sex-typing was revealed by our 422 community participants. Using Newcombe *et al.*'s agreement criterion, just six activities were sex-typed in our study compared to sixty-one activities in the earlier project.

Because Newcombe *et al.* had used just college undergraduates as a subject pool, we administered the sex-typing task to 119 college undergraduates

(58 women and 61 men). Twenty activities were sex-typed by this group. This was three times more sex-typing than was found for our mixed age group.

Results of the undergraduate student study, in particular, were different from what we had expected. We had hypothesized that because of laws requiring equal opportunity in the classroom, the much larger number of women entering the nation's work force, and the relaxation (at least in American culture) of traditional sex-role attitudes, younger subjects would pick up clues from their parents, teachers, peers, and the media and would hold less rigid attitudes about classifying any human activity in terms of its appropriateness for each sex. This hypothesis was not supported: generational differences existed in the social consensus of sex-typing with the youngest and oldest respondents sex-typing significantly more activities than did respondents of an intermediate age leading us to wonder why it is that young adults hold stronger gender-based beliefs about the 'appropriateness' of certain common everyday activities than do people of their parents' and grandparents' ages? Regardless of the reasons for their ratings, because younger subjects tend to have higher rates of sex-typing, we suggest that studies employing an undergraduate student sample only may not be representative of the populace as a whole.

Spatial and geographic testing

Males have been found consistently to out-perform females on several traditional psychometric tests of spatial ability, particularly those involving mental rotation. However, the evidence with respect to the relative performance of males and females on other spatial tasks and on more ecologically valid tasks such as wayfinding, map reading, and learning about environments is much less clear.

To compare the sexes on a wide array of tasks that more adequately represent what is meant by the terms 'spatial' and 'geographic', Montello *et al.* (1999) carried out seven groups of tests on a population of 79 persons (43 females and 36 males), aged nineteen to seventy-six years, with a mean age of forty-seven. Psychometric tests included the Hidden Patterns test (French *et al.*, 1963), Card Rotations test (French *et al.*, 1963), and the Vandenberg Mental Rotations test (Vandenberg and Kuse, 1978). These tests are composed of small abstract shapes drawn on paper. The tasks involve identifying a given shape when it has been variously rotated or otherwise transformed. A second group of tests focused on real world wayfinding or route learning. The route was approximately 420 metres long and wound around various buildings and other structures on a university campus. An initial guided walk encouraged incidental learning (i.e., no instructions were given except to follow the researcher). After the walk, participants were asked to sketch the route from memory. Two more walks were then completed during which participants knew they would be drawing a sketch map after each walk. Selected landmarks were pointed out and named by the instructor. A third class of tasks involved map learning. Two maps were provided, one of a hypothetical amusement park, the other a fictitious map of Grand Forks, North Dakota. Similar procedures to those undertaken in the campus route learning task were followed, including an incidental learning trial and two other intentional learning trials where instructions were given to remember the route and specific landmarks that were encountered. Sketch maps were drawn after each trial.

After the sketch maps were drawn, participants were asked to estimate straight-line distances between pairs of landmarks. The fourth set of tasks involved extant geographic knowledge of local, national, and international place locations. The tasks relied on participants' accumulated memory of information about features at all of these scales. For example, in the Santa Barbara area, magnitude estimates of distance between twenty-seven pairs of locations were obtained. Moving to a larger scale, relative locations of well-known world cities were examined with respect to a cardinal frame of reference (e.g.: Which is farther North: Washington, DC, or Paris, France?). Another task involved national knowledge of the US and focussed on city and state ordinal distance estimation. Questions here were of the type: Which city is closer to Denver, Colorado: Salt Lake City, Utah, or Omaha, Nebraska? Following this task, a city placement task was completed in which participants were given an outline map of the world and a list of fifteen cities with their countries identified to be placed appropriately on the map. Next came a set of object location experiments using a task developed by Silverman and Eals (1992). Here thirty-five items were placed on two table-tops, the walls above the tables, and the floor beneath the tables. After two minutes of studying this environment with instruction to memorize the identities and locations of everything they saw, participants were given a drawing of the environment and an alphabetical list of the thirty-five items that were displayed. The label of each item was to be placed at its equivalent location. The sixth class of tasks were verbal spatial descriptions. These were undertaken in conjunction with the campus route learning task. Participants were asked to describe the route they had just travelled as if telling someone else how to follow it. Finally, a set of self-report measures was collected including demographic variables, length of residence in the area, income level, education, and occupation. Prior familiarity with and exposure to the campus was measured, as was familiarity with the local city, and the various US and world cities represented in several of the above tasks. Participants also evaluated their spatial style, which included sense of direction, spatial ability, spatial preferences, and spatial anxiety scale.

Both univariate means tests and multivariate discriminate analysis performed on the data suggested a route-survey distinction between the sexes. Female superiority was observed on static object location memory tasks, whereas males were found to out-perform females on tests of newly acquired place knowledge derived from direct travel experience. No significant sex differences on tests of extant knowledge or map-derived knowledge were found.

Overall, a complex picture evolved, with males performing at a superior level on some tasks, females performing at a superior level on other tasks, and no significant differences based on sex being observed on other sets of tasks. For example, psychometric test results showed that males scored higher than females on the Mental Rotation tests, but not on the Hidden Patterns test, or on the Card Rotations test. This supports general evidence in the literature that the Vandenberg Mental Rotations test consistently shows higher male performance levels.

With respect to the route and survey tasks, we found that route distance error was significantly lower for males than for females, as was the case on a survey direction task. Females made fewer route landmark errors, and no male/female differences were observed for route turn errors and survey distance errors. No male/female differences in scores were found on the map use tasks, and no significant male/female difference was observed on the extant geographic knowledge task. Again, no significant differences were found in relative performance on the city cardinal location tasks, nor was any significant difference found on the task involving questions regarding which of a pair of cities was further north, south, east or west. In the replication of the Silverman and Eals (1992) experiment, males misplaced significantly more objects than did females, and males had significantly greater metric errors in estimating inter-point distances than did females. Results of the verbal directions tests showed that females referred to cardinal directions less frequently than did males, and they used more non-metric distance terms and fewer metric distance terms than did males.

Montello *et al*. (1999) found that some of the results of their study correlated well with the traditional findings in the literature (e.g. male superiority on the Mental Rotations test). Recent results indicating that females excelled at landmark memorization and localization whereas males excelled at survey type judgements also appear to be supported. However, the study also indicated that no sex differences were found on other standard psychometric tests, and females showed superiority on tests such as the Object Localization test (Silverman and Eals, 1992). In general, it was argued that the results provided sufficient evidence to give some support to a stylistic difference between the ways that males and females remember and use environmental information. This appeared to be a route-survey distinction in that males showed more accuracy in estimating metric distances and straight line directions, whereas females were more accurate at recalling object locations.

Cognitive mapping research in the future

In the previous pages we have tried to convey what is known about gender- and sex-related differences in cognitive mapping knowledge and spatial abilities. We have also attempted to illustrate tasks in which differences have and have not been found, and to flesh out just how much difference between the genders is reliably present.

A controversy in the literature which is almost as divisive as the scholars' quest to 'best' define the terms 'sex' and 'gender' is the question of whether it is appropriate to focus on difference. Pioneer feminists in the 1970s conducted research on women to expose societal inequalities and to suggest ways to eradicate them. The need for between sex research comparisons was expected to be short-lived. Such has not been the case.

Hyde (1994) and McDowell (1997), among others, urge feminist scholars not to abandon gender difference as a research area because in doing so they would lose their power to influence it and make it a non-sexist resource. Jones *et al.* (1997) suggest that 'feminist methodology should strive to make everyday life both a politically and practically important site of research' (p. xxx). This, too, is our message. We agree that yet another rash of studies pointing out known differences between women and men is not needed. It seems time to undertake a research agenda aimed at attempting to account for the prevalence of sex- and gender-related differences in specific spatial abilities and performance on spatial tasks over time. It is also time to recognize the many areas of spatial ability and behaviour where there is similarity between women's and men's performance and competence and to account for why this is so.

It makes sense to us to start by exploring the spatial experience of children in the late 1990s and into the new millennium. The significance of life-span preferences for certain highly spatial free-time activities (assembling things, tinkering, building models, designing clothes), educational courses, and sports needs to be looked at anew. It is important to recognize that preferences for education, activities, sports, and games are passed on generationally. Today's households are different from those of the past. We now have many more homes that can be characterized by the presence of single parents, teen parents, gay or lesbian parents, non-marital cohabitation, commuter marriages, dual-worker families, inter-generational families, and so on. We need scholarly studies to see how the gender division of labour in the home and parenting practices today may affect the sex-typing of human activity, and then evaluate the impact this stereotyping may have for the development of girls' and boys' cognitive maps. It is also essential to heed Sutton-Smith's (1986) observation that children's play is facing increased adult control and supervision and that home has increasingly become the major play space for children as well as adults. Studies are needed to discern whether indoor play fosters the development of spatial awareness and competence in the same way as outdoor play. How relevant is early spatial experience for the full development of one's cognitive map?

Few would dispute that an increasing number of highly educated, technology-oriented workers will dominate Western economies in the twenty-first century. A changing work force will be needed – one whose workers will regularly use computers and other forms of advanced technology on the job and be required for their work to travel and find their way around a diverse set of national, international, and virtual locales. Geographers have the ability to discern which cognitive mapping abilities are especially crucial for various employment fields and free-time endeavours. The value of new technologies (video games, CD ROMs, virtual reality) versus real world experience in fostering spatial awareness and cognitive-mapping competence also needs investigation, particularly because early evidence has indicated that computer-related applications may differentially advantage men (Colley *et al.*, 1994; Brosnan and Davidson, 1994).

As the world becomes smaller and boundaries between individual countries and regions of the world blur, it seems important to heed some findings from the Malinowski and Gillespie (1998) study. Remember that they found that participants from bilingual homes fared worse than participants from monolingual backgrounds in performing complex, real world wayfinding tasks. If this finding is replicated in future studies, it seems crucial that researchers explore how various cultural patterns and/or socio-economic influences in the home and in other arenas of life determine how spatial knowledge is created, communicated, validated, and valued. Geographers might also commit to undertake a critical assessment of the new geography curriculum recently introduced into US schools to be certain that it is meeting its aim to prepare all students geographically to participate fully in the new millennium.

Fausto-Sterling (1997) suggests that 'the development of scientifically sound theories about the evolution of human behavioural patterns and their relationship to contemporary behaviour could emerge from collaborations between social scientists, evolutionists, and behavioural biologists' (p. 242). We second this and others' calls for more multidisciplinary research teams and scholarly partnerships between women and men researchers to better understand cognitive mapping knowledge and abilities today in a gendered and sexed world. We heed Jones *et al.*'s (1997: xxx) advice that 'feminist methodology should help recover specificity among, rather than impose generality upon, research subjects'. More long-term, longitudinal, cross-cultural studies involving participants with varied abilities, knowledge, and expectations are needed to better understand complex spatial behaviour. We also hope to see encouragement given to the use, whenever appropriate, of qualitative research methodologies which engage researchers in social interaction with their research participants (Katz, 1994; Raghuram *et al.*, 1998).

We return to our original quote from Albert Einstein: 'If we knew what it was we were doing, it would not be called research, would it?' We are glad researchers looking at sex, gender, and cognitive mapping do not yet know exactly what they are doing. That is why this is such an interesting and sometimes perplexing area of research – but one ripe with options for future research studies as the world and the people in it undergo change.

Acknowledgements

The research reported in *Cognitive Mapping Research Today* was partially sponsored by grant #SB93–18643 from the National Science Foundation. We are grateful for this support.

References

Amponsah, B. and Krekling, S. (1997) Sex differences in visual-spatial performance among Ghanaian and Norwegian adults. *Journal of Cross-Cultural Psychology*, 28, 81–92.

Annett, M. (1985). *Left, Right, Hand, and Brain: The Right Shift Theory*. London: Erlbaum.

—— (1994) Handedness as a continuous variable with dextral shift: Sex, generation and family handedness in subgroups of left and right handers. *Behavioral Genetics*, 24, 51–63.

Beaumont, P., Gray, J., Moore, G. and Robinson, B. (1984) Orientation and wayfinding in the Tauranga departmental building: A focused post-occupancy evaluation. *Environmental Design Research Association Proceedings*, 15, 77–91.

Bem, S. (1981) *Bem Sex Role Inventory Professional Manual*. Palo Alto, CA.: Consulting Psychologists Press.

Berenbaum, S., Korman, K., and Leveroni, C. (1995) Early hormones and sex differences in cognitive abilities. *Learning and Individual Differences*, 7, 303–21.

Beyer, S. (1990) Gender differences in the accuracy of self-evaluations of performance. *Journal of Personality and Social Psychology*, 59, 960–70.

Born, M., Bleichrodt, N. and Van der Flier, H. (1987) Cross-cultural comparison of sex-related differences on intelligence tests: a meta-analysis. *Journal of Cross-Cultural Psychology*, 18, 283–314.

Braun, C. and Giroux, J. (1989) Arcade videogames: proxemic, cognitive and content analyses. *Journal of Leisure Research*, 21, 92–105.

Brosnan, M. (1998) The implications for academic attainment of perceived gender-appropriateness upon spatial task performance. *British Journal of Educational Psychology*, 68, 203–15.

—— and Davidson, M. (1994) Computerphobia: is it a particularly female phenomenon? *The Psychologist*, 7(2), 73–8.

Brown, L., Lahar, C. and Mosley, J. (1998) Age and gender-related differences in strategy use for route information: a 'map-present' direction-giving paradigm. *Environment and Behavior*, 30(2), 123–43.

Bryant, K. (1982) Personality correlates of sense of direction and geographical orientation. *Journal of Personality and Social Psychology*, 43, 1318–24.

Buss, D. (1995) Psychological sex differences: origins through sexual selection. *American Psychologist*, 50, 164–8.

Chagnon, N. (1977) *Yanomamo, the Fierce People*. New York: Holt, Rinehart and Winston.

Choi, J. and Silverman, I. (1997) Sex dimorphism in spatial behaviors: applications to route learning. *Evolution and Cognition*, 2, 165–71.

Colley, A., Gale, M. and Harris, T. (1994) Effects of gender role identity and experience on computer attitude components. *Journal of Educational Computing Research*, 10(2), 129–37.

Crawford, M., Chaffin, R. and Fitton, L. (1995) Cognition in social context. *Learning and Individual Differences*, 7(3), 341–62.

Dabbs, J., Chang, E-Lee, Strong, R. and Milun, R. (1998) Spatial ability, navigation strategy, and geographic knowledge among men and women. *Evolution and Human Behavior*, 19, 89–98.

Deaux, K. and Major, B. (1987) Putting gender into context: an interactive model of gender-related behavior. *Psychological Review*, 94, 369–89.

Devlin, A. and Bernstein, J. (1995) Interactive wayfinding: use of cues by men and women. *Journal of Environmental Psychology*, 15, 23–38.

Eals, M. and Silverman, I. (1994) The hunter-gatherer theory of spatial sex differences: proximate factors mediating the female advantage in recall of object arrays. *Ethology and Sociobiology*, 15, 95–105.

Fausto-Sterling, A. (1997) Beyond difference: a biologist's perspective. *Journal of Social Issues*, 53(2), 233–58.

Fischer, S., Hickey, D., Pellegrino, J. and Law, D. (1994) Strategic processing in dynamic spatial tasks. *Learning and Individual Differences*, 6, 65–105.

French, J., Ekstrom, R. and Price, L. (1963) *Kit of Reference Tests for Cognitive Factors*. Princeton, NJ: Educational Testing Services.

Galea, L. and Kimura, D. (1993) Sex differences in route-learning. *Personality and Individual Differences*, 14, 53–65.

Geary, D. (1995) Sexual selection and sex differences in spatial cognition. *Learning and Individual Differences*, 7, 289–301.

—— (1996) Sexual selection and sex differences in mathematical abilities. *Behavioral and Brain Sciences*, 19(2), 229–47.

Gittler, G. (1990) Three-dimensional cube comparison test (3DC). A Rasch-calibrated spatial ability test. *Theoretical Conception and Test Manual*. Weinheim, Germany: Beltz Test.

Golledge, R., Dougherty, V. and Bell, S. (1995) Acquiring spatial knowledge: survey versus route-based knowledge in unfamiliar environments. *Annals of the Association of American Geographers*, 85, 134–58.

—— Montello, D. and Self, C. 1993. Spatial competence: The contribution of socio-cultural and gender factors in measures of sex-related differences. Proposal submitted to National Science Foundation, Geography and Regional Science Program.

—— Ruggles, A., Pellegrino, J. and Gale, N. (1993) Integrating route knowledge in an unfamiliar neighborhood: along and across route experiments. *Journal of Environmental Psychology*, 13, 293–307.

Grimshaw, G., Sitarenios, G. and Finegan, J. (1995) Mental rotations at 7 years: Relations with prenatal testosterone levels and spatial play experiences. *Brain and Cognition*, 29, 85–100.

Guilford, J. and Zimmerman, W. (1956) *Guilford-Zimmerman Aptitude Survey*. Beverly Hills, CA: Sheridan.

Hall, E. (1966) *The Hidden Dimension*. Garden City, NJ: Doubleday.

Harris, L. (1981). Sex-related variations in spatial skill. In Liben, L.S., Patterson,

A.H. and Newcombe, N. (eds), *Spatial Representation and Behavior Across the Life Span*. New York: Academic, pp. 83–128.

Henrie, R., Aron, R., Nelson, B. and Poole, D. (1997) Gender-related knowledge variations within geography. *Sex Roles*, 36, 605–23.

Hines, M., McAdams, L., Chiu, L., Bentler, P. and Lipcamon, J. (1992). Cognition and the corpus callosum – verbal fluency, visuospatial ability, and language later-alization related to midsagittal surface areas of callosal subregions. *Behavioral Neuroscience*, 106(1), 3–14.

Holding, C. and Holding, D. (1989) Acquisition of route network knowledge by males and females. *The Journal of General Psychology*, 116, 29–41.

Huang, J. (1993) An investigation of gender differences in cognitive abilities among Chinese high school students. *Personality and Individual Differences*, 15(6), 717–19.

Hyde, J. (1994). Should psychologists study gender differences? Yes, with some guidelines. *Feminism and Psychology*, 4, 507–12.

Jackson, P. (1996) How will route guidance information affect cognitive maps. *Journal of Navigation*, 49(2), 178–86.

James, K. and Cropanzano, R. (1990) Perceived equity of a colleague's outcome: effects on performance. *Social Justice Research*, 4, 169–85.

—— and Greenberg, J. (1997) Beliefs about self and about gender groups: inter-active effects on the spatial performance of women. *Basic and Applied Social Psychology*, 19(4), 411–25.

Jones, J., Nast, H. and Roberts, S. (1997) *Thresholds in Feminist Geography*. Lanham, MD: Rowman and Littlefield.

Katz, C. (1994). Playing the field: questions of fieldwork in geography, *Professional Geographer*, 46(1), 67–72.

Kimura, D. and Hampson, E. (1994) Cognitive pattern in men and women is influenced by fluctuations in sex hormones. *Current Directions in Psychological Science*, 3, 57–61.

Kitchin, R. (1996) Are there sex differences in geographic knowledge and under-standing? *The Geographic Journal*, 162(3), 273–86.

Lawton, C. (1994) Gender differences in way-finding strategies: relationship to spatial ability and spatial anxiety. *Sex Roles*, 30, 765–79.

—— Charleston, S. and Zieles, A. (1996) Individual- and gender-related differ-ences in indoor wayfinding. *Environment and Behavior*, 28, 204–19.

Lott, B. (1997) The personal and social correlates of a gender difference ideology. *Journal of Social Issues*, 53(2), 279–98.

Malinowski, J. and Gillespie, W. (1998) Examining differences in performance on a real-world, large-scale wayfinding task. Paper presented at meeting, 94th Annual Meeting of the American Association of Geographers, 27 March 1998, at Boston, MA.

McBurney, D., Ganlin, S., Devineni, T. and Adams, C. (1997) Superior spatial memory of women – stronger evidence for the gathering hypothesis. *Evolution and Human Behavior*, 18(3), 165–74.

McDowell, L. (1997) Women/gender/feminisms: doing feminist geography. *Journal of Geography in Higher Education*, 21(3), 381–400.

McGee, M. (1982) Spatial abilities: the influence of genetic factors. In Potegal, M. (ed.), *Spatial Abilities: Development and Physiological Foundations*. New York: Academic.

McGivern, R., Mutter, K., Anderson, J., Wideman, G., Bodar, M. and Huston, P. (1998) Gender differences in incidental learning and visual recognition memory: support for a sex difference in unconscious environmental awareness. *Personality and Individual Differences*, 25, 223–32.

McGuiness, D. and Sparks, J. (1983) Cognitive style and cognitive maps: sex differences in representations of a familiar terrain. *Journal of Mental Imagery*, 7, 91–100.

Meehan, A. and Overton, W. (1986) Gender differences in expectancies for success and performance on Piagetian spatial tasks. *Merrill-Palmer Quarterly*, 32, 427–41.

Miller, L. and Santoni, V. (1986) Sex-differences in spatial abilities: strategic and experiential correlates. *Acta Psychologica*, 62(3), 225–35.

Moffat, S., Hampson, E. and Hatzipantelis, M. (1998) Navigation in a 'virtual' maze: sex differences and correlations with psychometric measures of spatial ability in humans. *Evolution and Human Behavior*, 19, 73–87.

Money, J., Alexander, D., and Walker, H., Jr. (1965) *A Standardized Road Map Test of Direction Sense*. Baltimore, MD: Johns Hopkins.

Montello, D., Lovelace, K., Golledge, R. and Self, C. (1999) Sex-related differences and similarities in geographic and environmental spatial abilities. *Annals of the Association of American Geographers*, 89(3), 515–534.

—— and Pick, H. (1993) Integrating knowledge of vertically aligned large-scale spaces. *Environment and Behavior*, 25(4), 457–84.

Newcombe, N., Bandura, M. and Taylor, D. (1983) Sex differences in spatial ability and spatial activities. *Sex Roles*, 9, 377–86.

Newell, A., and Simon, H. (1972) *Human Problem Solving*. Englewood Cliffs, NJ: Prentice Hall.

Nordvic, H. and Amponsah, B. (1998) Gender differences in spatial abilities and spatial activity among university students in an egalitarian educational system. *Sex Roles*, 38, 1009–23.

O'Laughlin, E. and Brubaker, B. (1998) Use of landmarks in cognitive mapping: gender differences in self report vs. performance. *Personality and Individual Differences*, 24(5), 595–601.

Okagaki, L. and Frensch, P. (1994) Effects of video game playing on measures of spatial performance: gender effects in late adolescence. *Journal of Applied Developmental Psychology*, 15, 33–54.

Passini, R., Proulx, G. and Rainville, C. (1990) The spatio-cognitive abilities of the visually impaired population. *Environment and Behavior*, 22, 91–118.

Piller, C. (1998) The gender gap goes high tech. *Los Angeles Times*, 25 August, A1.

Postma, A., Izendoorn, R. and De Haan, E. (1998). Sex differences in object location memory. *Brain and Cognition*, 36, 334–45.

Presson, C., DeLange, M. and Hazelrigg, M. (1989) Orientation specificity in spatial memory: What makes a path different from a map of the path? *Journal of Experimental Psychology: Learning, Memory, and Cognition*, 15, 887–97.

—— and Hazelrigg, M. (1984) Building spatial representation through primary and secondary learning. *Journal of Experimental Psychology*, 10, 716–22.

Raghuram, P., Madge, C. and Skelton, T. (1998). Feminist research methodologies and student projects in geography. *Journal of Geography in Higher Education*, 22(1), 35–48.

Richardson, J. (1994) Gender Differences in Mental Rotation, *Perceptual and Motor Skills*, 78, 435–48.

Roskos-Ewoldsen, B., McNamara, T., Shelton, A. and Carr, W. (1998) Mental representations of large and small spatial layouts are orientation dependent. *Journal of Experimental Psychology*, 24(1), 215–26.

Rovine, M. and Weisman, G. (1989) Sketch-map variables as predictors of wayfinding performance. *Journal of Environmental Psychology*, 9, 217–32.

Sadalla, E. and Montello, D. (1989) Remembering changes in direction. *Environment and Behavior*, 21, 346–63.

Schmitz, S. (1997) Gender-related strategies in environment development: effects of anxiety on wayfinding in and representation of a three-dimensional maze. *Journal of Environmental Psychology*, 17(3), 215–28.

Self, C., Golledge, R. and Montello, D. (1995) The gendering of spatial abilities. Paper presented at meeting, 91st Annual Meeting of the American Association of Geographers, 16 March 1995, Chicago, IL.

——, —— and —— (1997). Sport and recreational preferences of males and females: commonalities and differences. Paper presented at meeting, 93rd Annual Meeting of the American Association of Geographers, 4 April 1997, at Fort Worth, Texas.

——, ——, —— and Lovelace, K. (in progress). Generational differences in the sex-typing of spatial activities.

Signorella, M., Krupa, M., Jamison, W. and Lyons, N. (1986) A short version of a spatial activity questionnaire. *Sex Roles*, 14, 475–9.

Silverman, I. and Eals, M. (1992) Sex differences in spatial ability: Evolutionary theory and data. In Barkow, J.H., Cosmides, L. and Tooby, J. (eds), *The Adapted Mind: Evolutionary Psychology and the Generation of Culture*. New York: Oxford University Press, pp. 533–49.

—— Phillips, K. and Silverman, L. (1996) Homogeneity of effect sizes for sex across spatial tests and cultures: implications for hormone theories. *Brain and Cognition*, 31(1), 90–4.

Steele, C. (1997) A threat in the air: How stereotypes shape the intellectual identity and performance. *American Psychologist*, 52(6), 613–29.

Subrahmanyam, K. and Greenfield, P. (1994) Effect of video game practice on spatial skills in girls and boys. *Journal of Applied Developmental Psychology*, 15, 13–32.

Sutton-Smith, B. (1986) *Toys as Culture*. New York: Gardner.

Tanzer, N., Gittler, G. and Ellis, B. (1995) Cross-cultural validation of item complexity in a LLTM-calibrated spatial ability test. *European Journal of Psychological Assessment*, 11, 170–83.

Thurstone, L. and Thurstone, T. (1947) *Primary mental abilities*. Chicago, IL: Science Research Associates.

Tooby, J. and Cosmides, L. (1992) The psychological foundations of culture. In Barkow, J., Cosmides, L. and Tooby, J. (eds), *The Adapted Mind: Evolutionary Psychology and the Generation of Culture*. New York: Oxford University Press, pp. 19–136.

—— and DeVore, I. (1987) The reconstruction of hominid behavioral evolution through strategic modeling. In Kinzey, W. (ed.), *The Evolution of Human Behavior: Primate Models*. New York: SUNY, pp. 183–237.

Vandenberg, S. and Kuse, A. (1978) Mental rotations: a group test of three-dimensional spatial visualization. *Perceptual and Motor Skills*, 47, 599–604.

van Winsum, W., Alm, H., Schraagen, J. and Rothengatter, T. (1990) Laboratory and field studies on route representation and drivers' cognitive models of routes.

Drive project V1041 Generic Intelligent Drive Support Systems, Traffic Research Centre VSC, University of Groningen, Netherlands.

Ward, S., Newcombe, N. and Overton, W. (1986) Turn left at the church, or three miles north: A study of direction giving and sex differences. *Environment and Behavior*, 18, 192–213.

Whitley, B. (1983) Sex-role orientation and self-esteem: A critical meta-analytic review. *Journal of Personality and Social Psychology*, 44, 765–78.

Whitkin, H., Oltman, P., Raskin, E. and Karp, S. (1971) *A manual for the embedded figures test.* Palo Alto, CA: Consulting Psychologists Press.

13 Cognitive mapping without visual experience

Simon Ungar

Introduction

Try this simple experiment: close your eyes tightly, stand up, and walk to the other side of the room and back. You have just simulated for yourself what it is like to be blind. Well, not quite: there are several important factors missing. First, you knew all along that you could open your eyes at any minute if you ran into trouble (e.g., a large hard obstacle). A blind person does not have that option for recovering from a mistake. Second, you almost certainly used your visually derived mental 'map' of the room's layout to guide you. Think how much harder it would have been to do the same thing in an unfamiliar room. Thirdly, you drew on a set of spatial concepts and orientation skills developed across your life-span that involved vision as a major unifying sense; the very first time you, as an infant, watched your hand as you reached out for an object, you were already learning about space through vision.

Even this experiment hardly brings you close to the everyday experience of someone who is totally, congenitally blind. Such a person has no visual memories of particular spaces, and has had no direct *visual* input into the development of their spatial understanding in general. Their experience of space comes from hearing, touch and movement, and yet they can engage in pretty much all the activities that a sighted person can. How is this possible when we, as sighted people, place so much importance on visual experience in our lives? Through theories and research in cognitive psychology and behavioural geography, this chapter will explore the way(s) in which blind people experience and represent space. The first section will trace the history of thinking on the subject, the second will assess current work while the third will look to the future.

The past: getting the question right

What is it about vision that makes it so well suited for spatial representation? A number of features of the visual system make this modality appear to be better suited to spatial information than any other (Foulke and Hatlen,

1992; Millar, 1994; Thinus-Blanc and Gaunet, 1997). Vision provides relatively simultaneous perception of a large spatial field: although the point of foveation is quite limited, other objects are still present in peripheral vision as our attention wanders round a particular scene. In a sense, haptic exploration is like foveation without peripheral vision, in that the positions of objects not currently being attended to must be maintained in memory and no cues are available to draw attention in any particular direction. Vision is more precise than audition both in terms of accuracy of localization (distance and direction information) and identification of objects (features that tell us what something is). These advantages of vision for spatial perception have often led theorists to assume that blind and visually impaired people must necessarily be deficient in spatial abilities.

The spatial understanding of blind people became a topic of study in the late seventeenth century, when the English philosopher Locke attempted to answer a question put to him by his friend Molyneux:

> Suppose a man born blind, and now adult, and taught by his touch to distinguish between a cube and a sphere of the same metal, and nighly of the same bigness, so as to tell, when he felt one and the other, which is the cube, which the sphere. Suppose then the cube and sphere placed on a table, and the blind man be made to see: quaere, whether by his sight, before he touched them, he could now distinguish and tell which is the globe, which the cube?
>
> (Locke, *An Essay Concerning Human Understanding*)

Locke's answer to this question is negative. As an empiricist, he believed that our minds originate as a 'blank slate' (*tabula rasa*) and that all concepts we have are derived from our sensory experiences. Thus a congenitally blind person would only have *tactile* impressions of objects, and these could not automatically allow him to recognize the same objects by sight. It is only by integrating experiences from different senses that we build up abstract (amodal) concepts. In contrast to Locke, rationalist philosophers argued that abstract ideas about the world (including spatial concepts) are present at birth: we recognize something as spherical, in any sensory modality, by matching it to a pre-existing concept of a sphere.

The dichotomy of empiricism versus rationalism has continued to influence theories of the mind in the form of the famous 'nature/nurture debate'. However Millar argues that this way of thinking about the problem offers us little in either theoretical or practical terms (see Millar, 1994, for a fuller discussion of this issue), and she suggests an alternative reading of the question, focusing on the empirical issue of how perception relates to knowledge. In other words, if one perceptual modality (i.e. vision) is missing, what (if any) effect does this have on our knowledge of the world? Here, the question is phrased in terms of information processing and directs us to a study of the nature of information handled by the various modalities, the way this information is processed and the way the resulting representations guide or influence spatial behaviour.

Logically, three answers have been proposed for this question (Andrews, 1983; Fletcher, 1980): the lack of visual experience may result in a total lack of spatial understanding (the 'deficiency' theory); it may result in spatial abilities which are similar to, but necessarily less efficient than, those of sighted people (the 'inefficiency' theory); or it may result in abilities which are qualitatively different from, but functionally equivalent to, those of sighted people (the 'difference' theory).

The first of these positions is exemplified by the work of von Senden (1932) who argued that spatial concepts are impossible in people who have been blind from birth, and that visual experience during some early period is essential for even a minimal understanding of space. This strongly empiricist position is based on the assumption that vision is the sense through which all spatial representation is derived. More recent work has undermined this position, as Von Senden's methods have come under critical scrutiny and as empirical evidence has accumulated which goes against his position.

It has proved considerably more difficult to distinguish between the inefficiency and difference theories, with much of the empirical evidence supporting either theory. This is because most studies have simply focused on current spatial abilities (competence) rather than on potential, for instance by looking at gross performance on spatial tasks without considering the specific representations or strategies underlying performance. More recently, some authors (Millar, 1994; Thinus-Blanc and Gaunet, 1997; Ungar *et al.*, 1995b; 1996a; 1997a) have begun to focus explicitly on the strategies used to solve spatial tasks and the relationship between these and spatial performance. If the poor performance of blind groups on spatial tasks is due to a necessary limitation imposed by the lack of vision on the range of strategies they can use to code spatial relations, this would support the inefficiency theory. If on the other hand it can be shown that blind people potentially have access to a range of strategies, some supporting excellent performance, then this would favour an explanation in terms of difference. Indeed, Millar's (1994) approach challenges the very relevance of explanations phrased in terms of differences between blind and sighted people in the processing of spatial information. She places all the emphasis on the information that is available to people under particular task conditions (e.g., blind versus sighted task conditions). We will return to these more recent approaches after a review of the literature.

The present: empirical evidence and theoretical frameworks

Some important distinctions

Near space versus far (haptic vs. locomotor) space

In research on spatial cognition in blind and visually impaired people, a distinction is usually drawn between 'near' and 'far' space. The former relates to small-scale or manipulatory space: areas that can be explored without changing the location of the body. The latter relates to medium- or large-scale space: areas in which locomotion is required for exploration. Although the main focus of this chapter is on the latter, research on small-scale spatial tasks in blind children tells us a lot about the nature of spatial representation in general.

In the absence of vision this distinction is very important for the performance of spatial tasks. In small-scale space, where haptic exploration with the hands and arms is used, object locations can be represented relative to one's own body, providing a stable egocentric frame of reference. In large-scale space where exploration involves locomotion, the body must translate (i.e. change location), and egocentric reference frames become less reliable.

Millar (1994) points out that vision provides at least three advantages for the sighted traveller: the coincidence of body-centred and external reference frameworks during locomotion; the ability to look forwards and backwards along a route and thus integrate the locations of spatially separated landmarks; and in terms of prior knowledge about coding the relations of planes and surfaces. Thus, despite some logical similarities between blind and sighted people's experiences of large-scale space, there are some important functional differences also.

Early versus late onset

In considering the understanding of space by blind people it is important to make a clear distinction between people who have been blind since birth or early in life, and those who have lost their sight later and have therefore had some visual experience. Exactly how long a period of visual experience is necessary or at least valuable for spatial development? In general, the performance of later-blinded people on a range of spatial tasks is more similar to that of sighted people than of early-blind people. However, studies have employed widely varying cut-off points for the distinction between early- and late-blindness, ranging from a few months to three years of age. Although the results remain interesting for theoretical purposes, it will be necessary to be more precise about this factor in future research.

Memory versus inferential tasks

Another important distinction is between tasks that require participants to make a response based on a spatial relation that has been directly experienced, and tasks that require participants to infer a new relation based on their direct experience. The former simply requires some form of spatial coding, while the latter requires that a transformation be performed on the coded information. This distinction has frequently been used to test for differences in spatial coding particularly in large-scale space where performance in inferential tasks is generally more efficient and reliable when based on external coding, for instance an integrated or 'map-like' representation of a spatial layout.

On the table-top: coding strategies in haptic space

Coding spatial relations

In performing any spatial task, one has the option of coding the location of an object either by reference to one's own body and/or movements, or relative to some external framework (see Tversky, this volume, for a detailed discussion of this topic). For instance, I can determine the position of a cup on my desk either by its distance and direction from where I am sitting (i.e. by extending my arm by a certain amount in a particular direction relative to my body) or by its position relative to the layout on my desktop (e.g., between the computer and the lamp). Either method should allow me to reach for it accurately (all other things being equal). A number of studies have focused on the way blind and sighted people spontaneously code the locations of objects in small-scale space (for a detailed review of these studies, see Millar, 1994).

Generally, such studies show that people with little or no visual experience (congenitally and early blind) tend to code spatial relations in small-scale space by reference to their own body co-ordinates and/or their arm movements during exploration of the experimental space. According to Millar (1982), this is because the 'type and reliability of spatial information' (p. 72) available under blind conditions differs from that available with vision, and these differences in the quality of experience generally prompt early-blind children to organize spatial information by different coding strategies from those that tend to arise from visual experience.

Consider, for instance, the task of repeatedly locating a cup of tea placed in a constant position on a desk as you remain seated at the desk, also in a constant position. With vision, it may be more natural to code the cup's position relative to other objects on the desk. In the absence of vision, this strategy would involve locating the reference objects by touch each time you wanted to take a sip. It would be far more efficient in this case simply to encode the cup's position relative to your own body co-ordinates or according to a reliably reproducible series of arm movements. A similar everyday example comes when we are sitting in a stationary train, and an adjacent train starts to move off. Initially we feel, on the basis of visual information, that our own train is moving, but kinaesthetic information soon tells us the truth. Here, visual information is clearly *less* reliable than kinaesthetic information.

In this sense, strategies are seen by Millar as 'optional forms of coding' which differ in the types of information selected (e.g., relationships between locations in space or relation of locations relative to the body mid-line) and the coding heuristics appropriate for a particular type of information (e.g., external frame of reference, self-referent, movement). The strategies are optional in the sense of being interchangeable, although they are not necessarily equally reliable in a given context (e.g., the moving trains example). Visual experience prompts children to attend to external cues (e.g., the interrelationships between locations) and this is the case both for sighted children performing the tasks blindfold and for late-blinded children. Congenitally blind children tend to neglect such cues and thus adopt different strategies.

These strategies produce reliable performance in most small-scale tasks, the chief exception being those that involve spatial inference or mental rotation. In these cases the cognitive load involved (in calculating a new spatial relation or updating spatial relations after rotation) is greater with body-centred and movement strategies, although it is not impossible. If external cues are attended to and used, however, performance is generally improved. Such external cues are potentially available even in the absence of vision.

This is supported by a study (Ungar *et al.*, 1995b) in which blind and partially sighted children were asked to examine and then reproduce a layout of shapes either from the same location or at a new location, 90° rotated around the display. In the analysis, the children's exploration strategies were examined in relation to their performance in the reproduction task. The analysis showed that children who adopted a strategy that related objects to each other and to the frame of the display were less affected by the rotation than children who learned the objects' locations by simply touching each one repeatedly. It seems likely that these exploration strategies were correlated with particular coding strategies: children using the former strategy probably coded each object location by reference to the rest of the display while children using the latter strategy probably coded object positions relative to their body and/or arm movements. Similar results were found more recently by Gaunet *et al.* (1997).

External cues also become advantageous when the task is very complex. Ungar *et al.* (1997a) asked early-blind and sighted participants to learn a complex map of a fictional town, and then to reconstruct it from memory. Performance of the blind group was significantly poorer than that of the sighted group. However, the large individual differences within the blind group were accounted for by the various strategies used to explore the map and organize the information. Better performance was associated with strategies that used external reference frameworks, and it is likely that such strategies were used by participants who had more experience of small-scale spatial tasks, such as reading tactile maps or using the Optacon.

Summary

Spatial tasks at the small-scale (as we have defined it) are generally performed well by early-blind participants relative to late-blind and sighted groups (Millar, 1994; Thinus-Blanc and Gaunet, 1997). Lack of visual experience tends to give rise to body-centred and/or movement-based coding strategies as these are generally more reliable at this scale under blind conditions, and generally prove to be functionally equivalent to those used by sighted people (Klatzky *et al.*, 1995). Such strategies prove less effective when tasks require mental reorganization, mental rotation or spatial inference, or when tasks are very complex. However congenitally and early-blind participants do have the potential to use externally-based coding strategies which support good performance even on these tasks.

Wayfinding and cognitive maps: locomotor space

Methods of externalizing cognitive maps

A similar range of methods has been used to investigate the cognitive maps of blind people as has been used with sighted people (see Kitchin, 1996, and this volume, for a review). Sketch mapping is not widely used as blind people are generally unfamiliar with this medium, but the construction of models is used. More common have been direct (re)production of a route, distance or direction by walking, distance estimation or direction estimation using some kind of pointer. The relative reliability and validity of these measures for blind people has been considered in a number of studies (Haber *et al.*, 1993; Kitchin and Jacobson, 1997; Morsley, 1989).

Familiar environments

By looking at spatial knowledge of familiar environments, we can investigate the structure of cognitive maps constructed over time in relatively large areas. It is generally impractical to familiarize participants with very large and complex areas due the time consuming nature of this process.

Casey (1978) asked blind and sighted school children to produce a plan of their school campus using model buildings. He found that the blind participants as a group were less accurate than sighted participants, but that some individual blind participants were very accurate. Casey also found that the performance of blind participants was correlated with their level of independent mobility, although the direction of causation was not

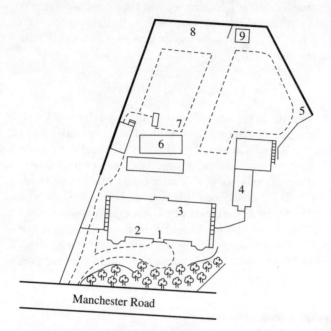

Figure 13.1 Layout of the school and grounds used to test children's knowledge of a familiar environment by Ungar *et al.* (1996b): 1 – Entrance steps, 2 – Dining room, 3 – Staff room, 4 – Assembly hall, 5 – Skittle alley, 6 – Play area, 7 – Sand pit, 8 – Boat, 9 – Swings.

explored. In an analysis of the models produced by Casey's participants, Golledge *et al.* (1996) pointed out that the congenitally blind participants had a tendency to linearize curved paths, that the maps were segmented and chunked rather than integrated and that features on familiar routes were more accurately represented than those on less familiar routes.

Ungar *et al.* (1996b) asked eighteen visually impaired children (aged 6 to 12.5 years) to estimate distances between nine locations around their school (see Figure 13.1). Rank order correlations were performed to compare the distance judgements of each age and visual status group with an accurate set of Euclidean and an accurate set of functional distances.[1] On the whole, children's relative distance judgements correlated more highly with the functional baseline than with the Euclidean baseline. In order to gain some impression of the mental representations underlying children's relative distance judgements, the data were analysed using multi-dimensional scaling. For all participants taken together a picture emerged in which functional distances were exaggerated, and these functional distances were based on habitual paths of movement by the children in their daily school activities. Overall the results were consistent with those obtained with adults by Rieser *et al.* (1980).

Constructed environments

One problem in testing people's knowledge of familiar environments is that it is impossible to control for individual differences in experience. Therefore a number of studies have tested children in novel environments – either an experimental environment constructed in the laboratory (Fletcher, 1980; Landau *et al.*, 1984; Rieser *et al.*, 1982; 1986) or an unfamiliar part of the real world (Dodds *et al.*, 1982; Espinosa *et al.*, 1998; Leonard and Newman, 1967; Ochaíta and Huertas, 1993)

Rieser *et al.* (1982; 1986) tested the ability of congenitally totally blind, later-blinded and blindfolded sighted adults to keep track of their position relative to a number of landmarks as they moved or imagined moving through an experimental layout of objects (see Figure 13.2). The participants learned the layout by walking with an experimenter from the start point to each of the landmarks in turn, returning to the start each time. The participants were then tested in two experimental conditions. In the locomotion condition, participants were led by a circuitous route to one of the experimental landmarks and asked to aim a pointer at each of the other landmarks in turn. In the imagination condition participants made pointer estimates from the start point but were asked to imagine that they were standing at one of the experimental landmarks.

The sighted and the adventitiously blind groups performed very accurately in the locomotion condition but less accurately in the imagination condition. In contrast, the performance of the early-blind group was similar in both conditions to the performance of the other groups in the imagination condition. Furthermore, the response latencies of the sighted and adventitiously blind were longer for the imagination condition than for the locomotion condition, whereas the latencies for the early-blind group in both conditions were similar to those of the other groups in the imagination condition.

Rieser *et al.* (1982) suggested that this pattern of results reflected differences in the way the task was performed. In the locomotion condition, the previous visual experience of the sighted and adventitiously blind groups afforded them a sensitivity to the changing perspective structure of the environment and thus allowed them to update their position automatically as they moved (cf. Gibson, 1986). In the imagination condition, without the locomotor information to support automatic updating, these groups had to resort to a strategy of calculating the relative positions of the landmarks. The early blind group, with similarly long latencies and high errors for the locomotion and the imagination conditions, appear to have used a calculation strategy in both conditions.

However, not all studies found poorer performance in early and congenitally blind participants. Loomis *et al.* (1993), in a replication of the Rieser *et al.* (1982; 1986) study, found no significant group differences in error scores and all but one of the early-blind participants performed at the level

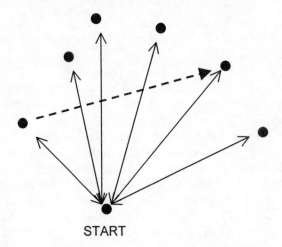

START

Figure 13.2 Layout used by Rieser, Guth and Hill (1982; 1986, based on a diagram in Thinus-Blanc and Gaunet, 1997). Solid arrows represent paths used to learn layout. Dotted line shows one example of an inferred path.

of the sighted participants. However response latencies of the early-blind participants were higher than those of the sighted participants, which might suggest a speed/accuracy trade off consistent with the use of a calculation strategy by the visually impaired participants. Moreover, Loomis *et al.* acknowledge that their participants had more experience of independent mobility that those of Rieser *et al.* (1982; 1986). Similarly, Klatzky *et al.* (1995) found no differences between congenitally blind, late-blind and sighted groups on a range of spatial tasks including those involving mental rotation and inference.

Hill *et al.* (1993) and Gaunet and Thinus-Blanc (1997) looked at the exploration strategies used by blind participants as they learned a large-scale layout of four objects in a large room similar to that of Rieser *et al.* (1982; 1986 see Figure 13.2). From observation of the participant's behaviour, each of these studies identified a number of strategies used to learn the experimental layouts; these are listed in Table 13.1. Both studies found a significant relationship between the pattern of strategy use and performance on tests of spatial knowledge of the experimental layout. Specifically, Hill *et al.* found that perimeter and gridline strategies used in isolation gave good knowledge of object location (indicating knowledge of the individual locations of objects), however in a test of integrated ('map-like') spatial knowledge of the layout, participants who used the perimeter strategy tended to perform poorly. Good performers tended to use object to object, perimeter to object or home-base to object strategies, and also used a wider range of strategies.

Table 13.1 Strategies identified in the studies by Hill *et al.* (1993) and Gaunet and Thinus-Blanc (1997)

Strategy	Description	Studies
Perimeter	Explored the boundaries of an area to identify the area's shape, size and key features around its perimeter, by walking along the edge of the layout	Hill *et al.*
Grid	Investigated the internal elements of an area to learn their spatial relationships, by taking straight-line paths from one side of the layout to the other	Hill *et al.*
Object to object	Moving repeatedly from one object to another, or feeling the relationship between objects using hand or cane	Hill *et al.*
Perimeter to object	Moving repeatedly between an object and the perimeter	Hill *et al.*
Home base to object	Moving repeatedly between the home base (origin point for exploration) and all the others in turn	Hill *et al.*
Cyclic	Each of the four objects visited in turn, and then returning to the first object	Gaunet and Thinus-Blanc
Back-and-forth	Moving repeatedly between two objects	Gaunet and Thinus-Blanc

Gaunet and Thinus-Blanc found that the cyclic patterns were used predominantly by early-blind participants, whereas late-blind and blind-folded sighted participants tended to use the back-and-forth strategy. Both within and across groups, use of a back-and-forth strategy was associated with good performance, whereas cyclic exploration was associated with poor performance, on a number of tests of spatial knowledge of the layout. The authors suggest that cyclic exploration formed the basis of a sequential representation of the layout whereas the back-and-forth strategy, like the object to object strategy of Hill *et al.*, formed the basis of a more integrated representation.

Novel environments

Although constructed environments allow us to test spatial performance independently of prior learning, practical considerations restrict these environments to a relatively small scale, typically within a large room. Exploration of such environments does involve locomotion, but the distances involved are very limited, and importantly are less than the normal range of humans. By testing participants in real environments with which they are unfamiliar, we gain the advantages of both familiar spaces (ecological validity) and of novel spaces control of experience. The cost is in terms of the practicality of familiarizing large numbers of participants with a large chunk of urban environment, and consequently relatively few studies of this kind have been carried out.

Dodds *et al*. (1982) introduced congenitally and late totally blind children (mean age: 11.5 years) to a short urban route by leading them along it four times. As they walked the route children were repeatedly asked to make pointer estimates to a number of locations along the route. Overall, errors in direction estimation increased with distance from the target, but this effect was considerably greater for the congenitally blind children, who were less accurate overall than the late-blind children. This finding suggests that visual experience facilitated the construction of co-ordinated spatial representations of locomotor displacements. As all the children were able to walk the route, Dodds *et al*. argued that the congenitally blind children must have coded the route in terms of body centred distances and changes of heading, but were not able to integrate this information into an externally-based representation of the layout. However, it is clear that this is not a necessary consequence of a lack of visual experience, as a number of congenitally blind participants performed at a similar level to sighted participants. Similar results were found by Espinosa *et al*. (1998) with blind adults in unfamiliar parts of central Madrid and suburban Sheffield.

In another study, Ocháita and Huertas (1993; Ocháita *et al*., 1991; Rosa and Ocháita, 1993) familiarised blind children and adolescents (from nine years to seventeen years) with a route linking seven landmarks in a real environment (school grounds or a public square) by leading them along it once. On three subsequent days, each participant led the experimenter along the route. At the end of each session, participants were asked to construct a scale model of the space and to estimate between-landmark distances. No differences were found between congenitally blind and late-blind groups on either measure.

Passini and Proulx (1988) asked blind and sighted adults to walk a route through two floors of a large, unfamiliar office building. After two guided walks, the participants were asked to walk the route unguided, while thinking aloud about the wayfinding decisions they made. The participants were then asked to produce a model of the route. The blind participants made significantly more wayfinding decisions and used more units of information (e.g., landmarks) than did the sighted group. However, both groups were equally able to produce models of the route.

Summary

With studies of large-scale space, the results have been less consistent. Most studies agree that early-blind participants perform as well as late-blind and sighted participants on tasks that involve spatial memory (e.g., reproducing angles or distances). However in tasks involving spatial inference (e.g., short-cutting, inferring crow-flight directions) the results are inconsistent: early-blind participants generally perform more poorly, but often no differences are found. Importantly, many studies have reported that some congenitally or early-blind individuals perform well within the range of sighted or late-blind groups. These individual differences indicate that the poor mean performance of blind groups does not reflect a necessary impairment in spatial ability resulting from lack of visual experience. Other kinds of experience and/or strategies used to solve the tasks are more likely to account for these individual differences.

Explaining the data

What, then, do these studies tell us about the spatial cognition of blind and visually impaired people? As we have already pointed out in the first section of this chapter, a strong 'deficiency' theory (e.g., von Senden 1932) will not do. In many of the studies cited above, congenitally totally blind participants were found to perform at the level of sighted participants on spatial tasks, including tests of spatial inference. We can confidently say that lack of visual experience does not prevent the acquisition of spatial representation.

As regards 'inefficiency' and 'difference' theories, it is less clear from the data which of these better characterizes the spatial cognition of blind people. This is partly because the studies themselves were often not designed to make this distinction. Poorer performance by congenitally blind groups versus later-blinded and sighted groups could be due to spatial processing and/or storage which is necessarily inferior in the absence of visual experience, or to the habitual use of different information processing strategies which produce poorer performance.

Two recent publications have argued convincingly for an interpretation in terms of difference. Both consider the poorer performance of congenitally blind groups on certain spatial tasks to be due to the use of different strategies: Millar (1994) emphasizes coding strategies while Thinus-Blanc and Gaunet (1997) focus on behavioural strategies.

Millar: informational conditions and spatial coding

In her recent book, Millar (1994; see also 1995 and 1997 for excellent summaries of her approach) proposes a new theory of spatial representation based on research with blind and sighted people. According to her approach, the study of spatial understanding and representation in the absence of visual experience tells us a great deal about spatial cognition in general, as well as providing practical solutions to the needs of blind adults and children. Central to her theory is her 'working model' of spatial development, called 'CAPIN' (convergent active processing in interrelated networks).

According to this, information from each of the different senses is special-ized but also complementary and overlapping, providing a significant degree of redundancy in the information entering the system.

Because of this overlap, spatial information is not the exclusive domain of one sensory modality. Spatially relevant information is available through senses other than vision (e.g., through hearing, touch, and movement) and this information can form the basis for spatial coding. However, the lack of one sense within the system tends to bias the way in which information is coded. While vision provides ready access to reliable information about external frames of reference (i.e. the relationship between external surfaces), touch, hearing, and movement do not. The most reliable forms of coding in blind conditions are those based on the body and on movement of the limbs. For this reason, congenitally blind children and adults generally tend to use egocentric coding in spatial tasks. While such coding strategies are most reliable for many spatial tasks, there are some tasks, e.g., those involving mental rotation or spatial inference, for which coding relative to external frameworks is advantageous.

The important point is that the processing of spatial information by congenitally blind people is not *necessarily* less efficient than in sighted people, as the 'inefficiency' theory proposes. It is misleading to focus purely on levels of efficiency rather than on the nature of coding used by partic-ipants. Congenitally totally blind people tend to code spatial relations egocentrically because this type of strategy generally works best for them. Moreover it is not the case, as Piaget and his colleagues (Piaget *et al.*, 1960) argued, that egocentric coding is simply an immature stage of development which is later entirely superseded by more logically rigorous forms of coding (e.g., Euclidean geometry). Millar points out that even very young children can use external frames of reference in certain task conditions, and that sighted adults often use egocentric frames of reference in situations where these are most reliable and efficient.

In large-scale space, Millar argues, this tendency to code spatial relations egocentrically results in a tendency to form sequential representations based on routes in contrast to the more global, map-like, externally-based repre-sentations characteristic of sighted people. However, it is stressed that the former means of coding is not necessarily inferior to the latter; it is simply more reliable under blind conditions in most cases. Millar points out that it is even possible to form map-like representations on the basis of route coding, although more cognitive effort is required to reach this level.

The information provided by the intact senses is potentially adequate to support other types of spatial coding (e.g., externally-based representations), provided blind children are systematically provided with the right kinds of cues for coding information by external frameworks. The practical task ahead, as Millar sees it, is to identify a range of methods for providing these cues. Possible interventions consistent with Millar's analysis are discussed in the next section.

Thinus-Blanc and Gaunet: behavioural strategies and performance

In their recent paper, Thinus-Blanc and Gaunet (1997) review a number of studies on the spatial performance of early- and late-blinded people, including studies on small-scale and large-scale environments. They point out that the main difference between early blind people on the one hand and sighted and late-blinded people on the other is in tests of spatial inference in locomotor space. In tasks where participants are asked to infer angles, distances or paths on the basis of a limited amount of spatial information, performance is generally better by late-blinded and sighted participants. However there are a number of revealing discrepancies in the data, which appear to be based on subtle variations in experimental factors and/or differences in the characteristics of participants.

The approach to explaining these discrepancies suggested by Thinus-Blanc and Gaunet, is to examine the ways in which people get to know their environment and solve spatial problems: in other words, the behavioural strategies spontaneously used by participants in spatial tasks. They define *strategy* as: 'a set of functional rules implemented by the participant at the various phases of information processing . . . [which] is assumed applicable to a wide range of situations' and allows the participant to 'reach an acceptable performance level without excessive cognitive effort' (p. 36). Although this differs from Millar's (e.g., 1994) use of the term (to mean a form of coding), these functional, behavioural strategies may well be associated with particular coding strategies, for instance a combination of a perimeter strategy, a perimeter to object strategy and an object to object strategy in the study by Hill *et al.* (1993) may be the behavioural counterpart of an external 'mapping' coding strategy, in which all objects are related to each other and to the outer frame of the room.

Thinus-Blanc and Gaunet propose three steps for such a research project. First, we must identify whether any behavioural regularities (e.g., different exploratory behaviours) can be consistently observed in both blind and sighted participants. Second, we need to identify which of these behaviours correlate with performance on various spatial tasks and thus merit the name 'strategy'. Finally a fine analysis of the strategies must be carried out to identify the actual cognitive mechanisms underlying them. This approach can potentially inform theory and generate a set of 'optimal' strategies which could form the basis of future training programmes for the poor performing blind participants.

This kind of approach has already been used to investigate the performance of blind and visually impaired people in small-scale (Gaunet *et al.*, 1997; Ungar *et al.*, 1995b; 1997a) and large-scale (Gaunet and Thinus-Blanc, 1997; Hill *et al.*, 1993; Tellevik, 1992) spatial tasks, and in tasks requiring transfer from small-scale to large-scale (Ungar *et al.*, 1996a).

Summary

In both the preceding explanations, differences in spatial performance by blind participants correspond to differences in behavioural and/or coding strategies used to acquire and organize spatial information. The fact that lack of visual experience tends to lead people to use particular strategies, accounts for the group differences observed in tests of spatial inference in locomotor space. It is a group *tendency* rather than an *inability* that gives rise to apparent differences at the group level.

Maximizing spatial potential

Education

Several practical implications for educators spring from Millar's (1994) analysis. The information about the structure of external space, which is so accessible in vision, must be substituted via touch and/or hearing. How this is to be done effectively is not straightforward. Millar stresses the importance of building on the information and coding strategies currently available to a blind child, and progressively integrating new sources of reference information with these existing ones; simply exposing a child to a new source of information which is rich in external cues (e.g., an electronic device, or an acoustically rich room) may not automatically cause the child to adopt a new coding system (Millar, 1994; 1995).

Similarly, training blind children or adults to use more effective behavioural strategies, as suggested by Thinus-Blanc and Gaunet (1997), may not automatically bring about changes in coding. For example, Ungar *et al.* (1995a) trained young blind and visually impaired children in strategies for learning a complex tactile map, based on the most effective strategies observed in a previous study (Ungar *et al.*, 1997a). After training, there was no change in participants' performance; the poorer participants appeared not to be able to apply the trained strategies and retained their habitual coding strategies. Such training perhaps needs to be more closely integrated with children's existing strategies and their understanding of space.

Methods for encouraging young blind children to explore independently may well facilitate their understanding of the structure of external space. Because a blind child is unable to detect environmental dangers like steps (even a small one is enough to cause a fall) and jagged edges, she may become reluctant to move out into space. Simply providing the child with a probe in the form of a cane or a wheeled toy may reduce this emotional barrier to movement (Morsley *et al.*, 1991; Pogrund *et al.*, 1993; Pogrund and Rosen, 1989).

Vision substitutes

Another approach has been to substitute vision with an electronic device that converts optical information about objects in the environment into auditory or tactile information. One such device, the Sonicguide™ (Kay, 1974) has been used in several studies with visually impaired infants and young children. It was hypothesized that providing children with auditory information about objects and surfaces in external space from an early age would facilitate their general understanding of the environment. In one study by Aitken and Bower (1982a; 1982b) three congenitally blind infants were given frequent sessions wearing the Sonicguide™ by their parents. The youngest of the three infants showed a number of spatially oriented behaviours (such as reaching and grasping) at approximately the appropriate age for sighted children whereas the other two infants apparently did not benefit from the Sonicguide™ at all. Warren (1994) provides a critical commentary on these and other, similarly inconclusive, studies.

In a new approach to sensory substitution, the Neural Rehabilitation Engineering Laboratory of the University of Louvain are integrating research in neuroscience, psychology and other fields to produce 'a model of the deprived sensory system connected to an inverse model of the substitutive sensory system' (PSVA, 1997). With recent advances in technology in the neurosciences, such a goal is increasingly realizable, and meshes well with Millar's (1994) recommendation to reintroduce the required redundancy into the system as a whole.

Tactile maps

Another potential tool for introducing blind and visually impaired people to the layout of the environment is a tactile map. A tactile map can provide a vicarious source of spatial information that preserves all the interrelationships between objects in space but presents those relationships within one or two hand-spans. The relevant information is presented clearly (irrelevant 'noise' which may be experienced in the actual environment, is excluded); with relative simultaneity (a map can be explored rapidly with two hands and with less demand on memory); and without other difficulties associated with travel in the real environment (e.g., veering or anxiety). Furthermore, if maps can encourage blind people to represent the environment by externally-based codes, they may form a crucial component of mobility training (Gilson *et al.*, 1965; Yngström, 1988).

It should be noted that tactile maps might have at least two important benefits. In a short-term sense, they can be employed to introduce a blind person to a particular space. However, the exercise of relating a map to the environment it represents, can potentially improve abstract level spatial thought in the long term, for instance encouraging the use of externally-based coding frameworks for structuring spatial representations of the environment by making the spatial relations between locations more accessible (Millar, 1994; 1995).

Espinosa *et al*. (1998) asked blind adults to learn a route through a novel environment either using a tactile map or by direct experience. Their knowledge of the route was tested by asking them to walk the route unguided (route knowledge) and to make direction estimates between locations on the route (inferred knowledge of the layout of the environment). Performance on both measures was best when participants learned the route by a combination of tactile map and direct experience. Participants who learned the route by direct experience alone performed poorly on both measures. Similar results were found in studies by Bentzen (1972) and by Brambring and Weber (1981), indicating that tactile maps can provide congenitally and early-blind people with a clearer impression of the spatial layout of an environment.

The evidence from the few studies with visually impaired children suggests that they have the potential to learn about, understand and use simple maps to perform orientation tasks in the environment.[2] However, until recently the evidence about visually impaired children's map use was based on small groups of older children (e.g., Gladstone, 1991) or single case studies (e.g., Landau, 1986). Therefore we carried out a number of experiments considering the potential of visually impaired children from five to twelve years to understand and use maps (Ungar *et al*., 1996a; 1997a; 1997b; Ungar *et al*., 1994).

In one study, we compared the performance of visually impaired children (aged from five to eleven years) who were asked to learn about an environment *either* by directly exploring that environment *or* by being shown a tactile map of it (Ungar *et al*., 1994). The environment consisted of a number of familiar toys arranged randomly around the floor of a large hall (see Figure 13.3). Tactile maps were constructed showing the location of all the toys. Both the totally blind and the partially sighted children were able to understand and use the maps. Most importantly, we found that the totally blind children learnt the environment more accurately from the map than from direct exploration. The results of this study demonstrated the importance of tactile maps for helping young totally blind children to form an impression of the space around them.

The general finding from our work is that young visually impaired children do have the potential to understand and use tactile maps. In some studies (e.g., Ungar *et al*., 1996a; Ungar *et al*., 1997a) it was found that the strategies children used to perform the map tasks affected their performance; visually impaired children who adopted effective tactile strategies often performed as well as or better than sighted and partially sighted children.

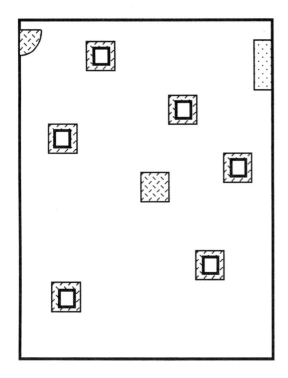

□   - Boxes containing
      toys

▨   - 1 metre squares
      of carpet

Figure 13.3 Example of the layout used by Ungar *et al.* (1994).

Summary

These studies indicate that even young blind children can acquire knowledge about the spatial structure of an environment from a tactile map, and
use this knowledge to make spatial decisions. Thus, tactile maps might be
an ideal means for emphasizing external frameworks in the environment,
which are not readily apprehended by direct experience alone. Spatial knowledge may be optimized when tactile maps are used in conjunction with
direct experience of the environment, supporting Millar's (1994; 1995)
recommendation that new strategies should be integrated with existing
ones. Individual differences in tactile map skill may be accounted for by
differences in strategies used to acquire information from the maps, and it
may be possible to train poorer map readers in more effective strategies.

Despite the optimistic findings of this research, tactile maps are still little used in practice, both in the classroom and in the outside world. Educators do not always have the time or resources to keep abreast of the latest developments in research. Furthermore, there is very little in the way of standards for the design and construction of tactile maps; practitioners make use of the resources they have to hand, and as a consequence, may find themselves re-inventing the wheel. Closer links are clearly needed between the research and practitioner communities.

Environmental modification and universal design

Recent developments in the sociology of disability and the rise of disabled peoples' movements have forced people researching impairment to reflect on the role of their work in relation to the lives of disabled people (Imrie, 1996; Kitchin, 1998; Oliver, 1990). In particular, a 'social model' of disability has been formulated which challenges the traditional, essentialist view of disability as a direct consequence of impairment. According to the social model, disability arises when environmental barriers (social, political or physical) prevent a person with impairments from functioning in society in the same way as able-bodied people. Focusing exclusively on the impairment can lead to solutions that attempt to fit the person to their environment, ignoring possibilities for adapting the environment to the person. The social model requires that disability be considered from the broadest possible perspective, where environmental modifications are preferred but intervention at the level of the individual is not ruled out.

Related to this, the principle of 'universal design' suggests that sensitive planning and design can yield environments that are equally suited to all people. This points us to environmental modifications which may benefit large numbers of people. Talking signs are an example of an environmental orientation aid which may be useful for blind and sighted people alike (Ungar, in preparation).

The future

In this final section, I will attempt to do two things: to hazard one or two guesses as to some likely directions of research in this area; and to highlight some areas which I think deserve attention in the future. This crystal ball gazing is necessarily based on my own personal experience and perspective within the field.

Early experiences: structuring the young child's spatial experiences

One implication of the research on coding strategies (Millar, 1994) and behavioural strategies (Thinus-Blanc and Gaunet, 1997; Ungar *et al.*, 1997a) is that early experience influences the way blind children process spatial information. Even at the earliest school age, children with similar levels of visual impairment already differ in their performance on spatial tasks. This leads to an emphasis on how the early environment and activities of blind infants are structured: in the absence of the structuring effect of vision, how do infants learn about the space around them?

Few studies have looked specifically at sources of spatial information in the environments of blind infants. In a review of the sparse literature, Warren (1994) identifies aspects of both the physical and social environment that may influence subsequent cognitive development. In this sense the opportunity to explore the properties and layout of the physical environment safely and the encouragement to do so may both be important.

Nielsen (1991) developed a 'little room' in which blind infants could explore a range of objects without the distracting and unpredictable noises of larger environments. She found that infants who were systematically exposed to this restricted environment exhibited relatively precocious search behaviour.

A number of studies on the social development of blind children (Als *et al.*, 1980; Pérez and Castro, 1994; Preisler, 1991; Rattray and Zeedyk, 1995; Rowland, 1983; Urwin, 1978) suggest that the joint activities of infant and care-giver are very different when the infant is blind. Care-givers often do not naturally compensate for their infant's impairment, and must learn to read the different kinds of social cues that the infant gives and respond to these. Likewise, it is possible that care-givers do not structure their interactions with their infant in a way that would optimize his/her experience of space. This aspect of early development has not received sufficient attention.

The 'neglected' visually impaired

While the literature on the effects of total blindness increases rapidly, there has been little research into the effects of differing degrees and types of visual impairment on spatial cognition. This is partly due to the historical significance of the Molyneux question (discussed earlier) which gives a more general theoretical importance to the study of blindness and space, but also to the increased complexity of designing studies which control for the varying profile of visual impairment across the population.

In one study that attempted to address these broader questions, Rieser *et al.* (1992) asked blind adults with varying degrees and types of residual vision to make direction and distance estimates in a familiar area. For the participants with early onset of impairment, they found that degree of field loss, but not degree of acuity loss, was negatively associated with accuracy of estimates. This result qualifies the finding (e.g., Rieser *et al.*, 1982; 1986) that lack of visual experience *per se* leads to difficulties in acquiring integrated representations of locomotor space. Here some forms of visual impairment led to difficulties while others did not. This result has major implications for provision of education and services for people with visual impairments and deserves to be a focus of research in the future.

Neuropsychology

With recent advances in techniques for scanning brain activity, a number of studies has considered the possibility that differences in processing spatial information between blind and sighted people may be evident at the level of brain function. In other words, lack of early visual experience may lead to different parts of the brain being used for spatial tasks. Studies using positron emission tomography (PET scan) (Veraart *et al.*, 1990; Wanet-Defalque *et al.*, 1988) and electroencephalogram (EEG) (Röder *et al.*, 1997), for instance, have found that activity in the occipital area of the cortex, usually used for processing visual signals, is generally higher than in sighted and late-blind people. This generalized increased activity is unlikely to support any specific brain function and is probably the result of surviving synaptic connections which normally disappear as a result of visual experience (Thinus-Blanc and Gaunet, 1997).

Another study using PET (Catalan-Ahumada *et al.*, 1993) found similarly raised levels of occipital activity during a spatial localization task, but also detected lower activity in the Parietal Area 7 of early blind participants. As Parietal Area 7 has been associated with spatial processing, the authors speculated that the difficulties with spatial tasks may be based in reduced function in this area.

Further developments in this field may well reveal further differences in localized brain function between early blind and sighted or late blind people. As our understanding of brain function increases, this may lead to new ways of substituting sensory information for early blind people. For instance, the Neural Rehabilitation Engineering Laboratory of the University of Louvain is simultaneously investigating differences in brain function and developing methods of sensory substitution based on this research (PSVA, 1997).

Cognitive maps and urban design

To some extent turning cognitive mapping research on its head, Vujakovic and Matthews (Matthews and Vujakovic, 1995; Vujakovic and Matthews, 1994) examined the extent to which differences in cognitive maps of special populations (in their case wheelchair users) reflect their experiences of the built environment. In their study, Vujakovic and Matthews asked geography undergraduates and wheelchair users working together in pairs to produce maps of the city centre of Coventry which reflected the wheelchair users' experiences of mobility in that area. In particular, one map showed the causes of specific mobility blackspots while another revealed the uneven 'mobility surface' of the area, where streets were graded according to their level of accessibility. Through the process of externalizing their cognitive maps of the area, the wheelchair users could contribute to the mapping process and at the same time make a political statement about the level of service provision to disabled people in Coventry.

The emphasis here is on the structure of the built environment itself rather than on the representation or processing of this structure. Rather than treating the cognitive maps of their participants as distortions of reality (Golledge, 1993), the maps were considered to be versions of reality reflecting a pressing need for environmental modification. The cognitive maps thus take on a political significance over and above their psychological interest. This method could usefully be applied to blind and visually impaired people (as indeed to many other minority populations). The added spatial aspect of this research can enrich previous questionnaire based studies of environmental barriers for blind and visually impaired people (Passini *et al.*, 1985; Passini *et al.*, 1986).

GIS and GPS

Another technological development may reduce or obviate the need for cognitive maps. The combination of geographical information systems (GIS) and global positioning systems (GPS) allows us to pinpoint our location quite precisely, and can give us detailed instructions about how to get to our destination. Such systems are already in use for in-car navigation and are being developed for use by blind people (Golledge *et al.*, 1991; Jacobson, 1994). Golledge and his colleagues (Golledge *et al.*, 1991; Golledge *et al.*, 1998) have been developing a portable personal guidance system which can provide a blind traveller with constantly updated information about their position and heading, about local objects, and can provide information about optimal routes. A major advantage of this technology is that the GIS component can be updated regularly providing the user with up-to-the-minute information about the accessibility of different routes within the environment. It seems likely that the ever-reducing cost and size of such technology will bring it within the reach of most blind people in the not-too-distant future. The effect this will have on their spatial understanding will be another interesting research question for the future.

Acknowledgements

I am grateful to Susanna Millar for her valuable comments on an earlier draft of this chapter. I am always indebted to the people who participated in my research and to my collaborators, in particular Mark Blades, Christopher Spencer, Angeles Espinosa, Esperanza Ochaíta, and Gordon Liebschner, for their input. Some of the work reported here was financially supported by the Economic and Social Research Council.

Notes

1   Euclidean distances are direct or 'crow-flight' distances. Functional distances are actual travelling distances or 'city block' distances.
2   It has been suggested that young children cannot use maps, as they lack an understanding of symbolic systems. Specifically they lack an understanding that a map 'stands for' or represents the environment (e.g. Liben and Downs, 1989; Millar, 1994). However, research has shown that even pre-school sighted children can understand and use simple maps for simple tasks (Blades, 1994). These children can make use of the 'correspondences' between the map and the environment (e.g. to make appropriate spatial responses) even if they do not yet fully understand how the map functions as a representation of the environment. See Uttal & Tan (this volume) for a more detailed discussion of this issue.

References

Aitken, S. and Bower, T.G.R. (1982a) Intersensory substitution in the blind. *Journal of Experimental Child Psychology*, 33, 309–23.

—— and —— (1982b) The use of the Sonicguide in infancy. *Journal of Visual Impairment and Blindness*, 76, 91–100.

Als, H., Tronick, E. and Brazelton, T.B. (1980) Stages of early behavioural organization: The study of a sighted infant and a blind infant in interaction with their mothers. In Field, T.M. (ed.), *High-Risk Infants and Children, Adult and Peer Interactions*. New York: Academic.

Andrews, S.K. (1983) Spatial cognition through tactual maps. Paper presented at the 1st International Symposium on Maps and Graphics for the Visually Handicapped, Washington, DC.

Bentzen, B.L. (1972) Production and testing of an orientation and travel map for visually handicapped persons. *New Outlook*, 66, 249–55.

Blades, M. and Spencer, C. (1994) The development of children's ability to use spatial representations. In Reese, H.W. (ed.), *Advances in Child Development and Behaviour*, 25. New York: Academic, pp. 157–99.

Brambring, M. and Weber, C. (1981) Taktile, verbale und motorische informationen zur geographischen orientierung blinder. (Tactile, verbal and motor information on the geographic orientation of the blind). *Zeitschrift für Experimentelle und Angewandte Psychologie*, 28, 23–37.

Casey, S. (1978) Cognitive mapping by the blind. *Journal of Visual Impairment and Blindness*, 72, 297–301.

Catalan-Ahumada, M., De Volder, A.G., Melin, J., Crucq, B. and Veraart, C. (1993) Increased glucose utilization in the visual cortex of blind subjects using an ultrasonic device. *Archives Internes de Physiologie et de Biochimie*, 101, 29.

Dodds, A.G., Howarth, C.I., and Carter, D.C. (1982) The mental maps of the blind: the role of previous experience. *Journal of Visual Impairment and Blindness*, 76, 5–12.

Espinosa, M.-A., Ungar, S., Ochaíta, E., Blades, M. and Spencer, C. (1998) Comparing methods for introducing blind and visually impaired people to unfamiliar urban environments. *Journal of Environmental Psychology*, 18, 277–87.

Fletcher, J.F. (1980) Spatial representation in blind children 1: development compared to sighted children. *Journal of Visual Impairment and Blindness*, 74, 318–85.

Foulke, E. and Hatlen, P.H. (1992). A collaboration of two technologies. Part 1: Perceptual and cognitive processes: their implications for visually impaired persons. *British Journal of Visual Impairment*, 10, 43–6.

Gaunet, F., Martinez, J.-L. and Thinus-Blanc, C. (1997) Early-blind subjects' spatial representation of manipulatory space: exploratory strategies and reaction to change. *Perception*, 26, 345–66.

—— and Thinus-Blanc, C. (1997) Early-blind subjects' spatial abilities in the locomotor space: exploratory strategies and reaction-to-change performance. *Perception*, 25, 967–81.

Gibson, J.J. (1986) *The Ecological Approach to Visual Perception*. Hillsdale, NJ: Erlbaum.

Gilson, C., Wurzburger, B. and Johnson, D.E. (1965) The use of the raised map in teaching mobility to blind children. *New Outlook*, 59, 59–62.

Gladstone, M. (1991) Spatial cognition and mapping abilities: a comparison of sighted and congenitally blind individuals. Unpublished honours dissertation, University of Edinburgh.

Golledge, R.G. (1993) Geography and the disabled – a survey with special reference to vision impaired and blind populations. *Transactions of the Institute of British Geographers*, 18, 63–85.

——, Klatzky, R.L. and Loomis, J.M. (1996) Cognitive mapping and wayfinding by adults without vision. In Portugali, J. (ed.), *The Construction of Cognitive Maps*. Dordrecht: Kluwer.

——, ——, ——, Speigle, J. and Tietz, J. (1998) A geographical information system for a GPS based personal guidance system. *International Journal of Geographical Information Science*, 12, 727–49.

—— Loomis, J.M., Klatzky, R.L., Flury, A. and Yang, X.L. (1991) Designing a personal guidance system to aid navigation without sight: Progress on the GIS component. *International Journal of Geographical Information Systems*, 5, 373–95.

Haber, L., Haber, R.N., Penningroth, S., Novak, K. and Radgowski, H. (1993) Comparison of 9 methods of indicating the direction to objects – data from blind adults. *Perception*, 22, 35–47.

Hill, E.W., Rieser, J.J., Hill, M.M., Hill, M., Halpin, J. and Halpin, R. (1993) How persons with visual impairments explore novel spaces – strategies of good and poor performers. *Journal of Visual Impairment and Blindness*, 87, 295–301.

Imrie, R. (1996). *Disability and the City: International Perspectives*. London: Paul Chapman.

Jacobson, R.D. (1994) GIS and the visually disabled – the spatial contribution to mobility. *Mapping Awareness*, July, 34–6.

Kay, L. (1974) A sonar aid to enhance spatial perception of the blind: engineering design and evaluation. *The Radio and Electronic Engineer*, 44, 40–62.

Kitchin, R. (1996) Methodological convergence in cognitive mapping research: investigating configurational knowledge. *Journal of Environmental Psychology*, 16, 163–85.

—— (1998) 'Out of place', 'knowing one's place': Towards a spatialised theory of disability and social exclusion. *Disability and Society*, 13, 343–56.

—— and Jacobson, R.D. (1997) Techniques to collect and analyze the cognitive map knowledge of persons with visual impairment or blindness: issues of validity. *Journal of Visual Impairment and Blindness*, 91, 393–400.

Klatzky, R.L., Golledge, R.G., Loomis, J.M., Cicinelli, J.G. and Pellegrino, J.W. (1995) Performance of blind and sighted persons on spatial tasks. *Journal of Visual Impairment and Blindness*, 89, 70–82.

Landau, B., Spelke, E. and Gleitman, H. (1984) Spatial knowledge in a young blind child. *Cognition*, 16, 225–60.

Leonard, J.A. and Newman, R.C. (1967) Spatial orientation in the blind. *Nature*, 215, 1413–14.

Liben, L.S. and Downs, R.M. (1989) Understanding maps as symbols: the development of map concepts in children. In Reese, H. (ed.), *Advances in Childhood Development and Behaviour*, 22. New York: Academic, pp. 145–201.

Loomis, J.M., Klatzky, R.L., Golledge, R.G., Cicinelli, J.G., Pellegrino, J.W. and Fry, P.A. (1993) Nonvisual navigation by blind and sighted – assessment of path integration ability. *Journal of Experimental Psychology – General*, 122, 73–91.

Matthews, M.H. and Vujakovic, P. (1995) Private worlds and public places – mapping the environmental values of wheelchair users. *Environment and Planning A*, 27, 1069–83.

Millar, S. (1982) The problem of imagery and spatial development in the blind. In de Gelder, B. (ed.), *Knowledge and Representation*. London: Routledge and Kegan Paul, pp. 111–20.

—— (1994). *Understanding and Representing Space: Theory and Evidence from Studies with Blind and Sighted Children*. Oxford: Oxford University Press.

—— (1995) Understanding and representing spatial information. *British Journal of Visual Impairment*, 13, 8–11.

—— (1997) Theory, experiment and practical application in research on visual impairment. *European Journal of Psychology of Education*, 12, 415–30.

Morsley, K. (1989) Enhancing the orientation and mobility of the visually impaired child: An evaluation of current theories and practices. Unpublished PhD thesis, Sheffield University.

—— Spencer, C. and Baybutt, K. (1991) Two techniques for encouraging movement and exploration in the visually impaired child. *British Journal of Visual Impairment*, 9, 75–8.

Nielsen, L. (1991) Spatial relations in congenitally blind infants: a study. *Journal of Visual Impairment and Blindness*, 85, 11–16.

Ochaíta, E. and Huertas, J.A. (1993) Spatial representation by persons who are blind: a study of the effects of learning and development. *Journal of Visual Impairment and Blindness*, 87, 37–41.

—— , —— and Espinosa, A. (1991). Representación espacial en los niños coegos: una investigación sobre las principales variables que la determinan y los procedimientos de objetivación más adecuados. [Spatial representation in blind children: an investigation into the main determining factors and the most appropriate externalization techniques]. *Infancia y Aprendizaje*, 54, 53–79.

Oliver, M. (1990) *The Politics of Disablement*. Basingstoke: Macmillan.

Passini, R., Delisle, J., Langlois, C. and Proulx, G. (1985) *Etude Descriptive de la Mobilite et de l'Orientation Spatiales chez des Handicapes Visuels en Milieu Urbain* (A descriptive study of the mobility and spatial orientation of the visually handicapped in the urban environment): Université de Montreal, Faculté de l'Amenagement.

—— Dupré, A. and Langois, C. (1986) Spatial mobility of the visually handicapped active person: a descriptive study. *Journal of Visual Impairment and Blindness*, 80, 904–7.

—— and Proulx, G. (1988) Wayfinding without vision: an experiment with congenitally, totally blind people. *Environment and Behaviour*, 20, 227–52.

Pérez Pereira, M. and Castro, J. (1994) *El Desarollo Psicológico de los Niños Ciegos en la Primera Infancia*. Madrid: Paidós.

Piaget, J., Inhelder, B. and Szeminska, A. (1960) *The Child's Conception of Geometry*. London: Routledge and Kegan Paul.

Pogrund, R.L., Fazzi, D.L. and Schreier, E.M. (1993). Development of a preschool 'Kiddy Cane'. *Journal of Visual Impairment and Blindness*, 87, 52–4.

—— and Rosen, S.J. (1989) The preschool blind child can be a cane user. *Journal of Visual Impairment and Blindness*, 83, 431–9.

Preisler, G.M. (1991). Early patterns of interaction between blind infants and their sighted mothers. *Child: Care, Health and Development*, 17, 65–90.

PSVA – Prosthesis for Substitution of Vision by Audition, [on-line]. (1997) Neural Rehabilitation Engineering Laboratory of the Catholic University of Louvain. Available: http://www.md.ucl.ac.be/entites/fsio/gren/Projets/PSVA.html [21st January 1999]

Rattray, J. and Zeedyk, M.S. (1995) Early interaction of visually impaired infants and their mothers. Paper presented at the British Psychological Society Developmental Psychology Section Annual Conference, University of Strathclyde.

Rowland, C.M. (1983) Patterns of interaction between three blind infants and their mothers. In Mills, A.E. (ed.), *Language Acquisition in the Blind Child: Normal and Deficient*. London: Croom Helm.

Rieser, J.J., Guth, D.A. and Hill, E.W. (1982) Mental processes mediating independent travel: implications for orientation and mobility. *Journal of Visual Impairment and Blindness*, 76, 213–18.

—— , —— and —— (1986) Sensitivity to perspective structure while walking without vision. *Perception*, 15, 173–88.

—— Lockman, J.J. and Pick, H.L. (1980) The role of visual experience in knowledge of spatial layout. *Perception and Psychophysics*, 28, 185–90.

—— Talor, C.R., Rosen, S., Hill, E.W. and Bradfield, A. (1992) Visual experience, visual field size, and the development of nonvisual sensitivity to the spatial structure of outdoor neighborhoods explored by walking. *Journal of Experimental Psychology – General*, 121, 210–21.

Röder, B., Rösler, F. and Henninghausen, E. (1997) Different cortical activation patterns in blind and sighted humans during encoding and transformation of haptic images. *Psychophysiology*, 34, 292–307.

Rosa, A., and Ochaíta, E. (eds), (1993) *Psicología de la Ceguera*. Madrid: Alianza Psicología.

Tellevik, J.M. (1992) Influence of spatial exploration patterns on cognitive mapping by blindfolded sighted persons. *Journal of Visual Impairment and Blindness*, 86, 221–4.

Thinus-Blanc, C. and Gaunet, F. (1997) Representation of space in blind persons: vision as a spatial sense? *Psychological Bulletin*, 121, 20–42.

Ungar, S. (in preparation). A city for different people: from exclusion to universal design.

—— , Blades, M. and Spencer, C. (1995a) The effectiveness of training visually impaired children in strategies for exploring tactile maps. Paper presented at the British Psychological Society Developmental Psychology Section Annual Conference, University of Strathclyde.

—— , —— and —— (1995b) Mental rotation of a tactile layout by young visually impaired children. *Perception*, 24, 891–900.

—— , —— and —— (1996a) The ability of visually impaired children to locate themselves on a tactile map. *Journal of Visual Impairment and Blindness*, 90, 526–35.

—— , —— and —— (1996b) The construction of cognitive maps by children with visual impairments. In Portugali, J. (ed.), *The Construction of Cognitive Maps*. Dordrecht: Kluwer, pp. 247–73.

—— , —— and —— (1997a) Strategies for knowledge acquisition from cartographic maps by blind and visually impaired adults. *The Cartographic Journal*, 34, 93–110.

—— , —— and —— (1997b) Teaching visually impaired children to make distance judgements from a tactile map. *Journal of Visual Impairment and Blindness*, 91, 163–74.

—— , —— , —— and Morsley, K. (1994) Can visually impaired children use tactile maps to estimate directions? *Journal of Visual Impairment and Blindness*, 88, 221–33.

Urwin, C. (1978) The development of communication between blind infants and their mothers. In Lock, A. (ed.), *Action, Gesture and Symbol: The Emergence of Language*. New York: Academic.

Veraart, C., Volder, A.G.D., Bol, A., Michel, C. and Goffinet, A.M. (1990) Glucose utilization in human visual cortex is abnormally elevated in blindness of early onset but decreased in blindness of late onset. *Brain Research*, 510, 115–21.

von Senden, S. M. (1932) *Space and Sight: The Perception of Space and Shape in the Congenitally Blind Before and After Operation*. Glencoe, IL: Free Press.

Vujakovic, P. and Matthews, M.H. (1994) Contorted, folded, torn: environmental values, cartographic representation and the politics of disability. *Disability and Society*, 9, 359–74.

Wanet-Defalque, M.-C., Veraart, C., Volder, A.D., Metz, R., Michel, C., Dooms, G. and Goffinet, A. (1988) High metabolic activity in the visual cortex of early blind human subjects. *Brain Research*, 446, 369–73.

Warren, D.H. (1994) *Blindness and Children: An Individual Differences Approach*. Cambridge: Cambridge University Press.

Yngström, A. (1988) The tactile map: the surrounding world in miniature. Paper presented at the 2nd International Symposium on Maps and Graphics for Visually Handicapped People, Nottingham: University of Nottingham.

14 The future of cognitive mapping research

Rob Kitchin and Scott Freundschuh

Introduction

As detailed in the chapters in this book, cognitive mapping research has developed over the past forty years into a vibrant and multidisciplinary field of study, with several discernible sub-fields. Whilst the studies that compose the body of cognitive mapping research provide both breadth and depth, it is clear that there are still many facets of spatial knowledge that remain unexamined or are need of further investigation. Indeed, the volume of research within each sub-field is highly uneven, with some focuses receiving a disproportionate amount of attention. Moreover, most sub-fields are characterized by a set of divergent and competing findings and theories, each seeking to adequately explain how we learn, store, process and use spatial knowledge. Each of the contributors to this volume detailed a specific future agenda to address questions so far left unexplored or inadequately answered. In this final chapter we collate, cross-reference and add to their suggestions for future research to provide a comprehensive agenda that will help guide cognitive mapping research as we enter the new millennium. We have divided our discussion into three main sections: theoretical, methodological, application.

Theoretical

Basic research

Whilst there has been significant progress concerning the investigation of basic research questions, it clear that more research is needed before we have a comprehensive understanding of all aspects of cognitive mapping. As argued by Tversky (Chapter 3), basic questions concerning levels and types of spatial knowledge, and how knowledge is structured (alignment, reference points, frames of reference, hierarchical organization, canonical axes), need more attention, as do questions relating to how knowledge is acquired from different sources (e.g., direct experience, maps, virtual reality) and whether such knowledge is treated in the same or different manner.

Research to date provides a number of competing theoretical models, each with fairly limited empirical support, and current comprehension needs to be supplemented by further examination. In particular, basic research is needed in regard to specific media of spatial learning, processes of integration, socio-temporal effects, spatial language, and the links between spatial thought and spatial behaviour.

Media of spatial learning

As detailed in Chapters 5 to 7 there have been significant advances in our understanding of how spatial information is learnt from a variety of media. This understanding, however, is far from complete. In relation to spatial learning through direct experience a number of key questions still remain in relation to the exact mode of learning. For example, how is perceptual data processed and turned into long-term memory, and how is such memory used to guide spatial behaviour? Further, what has to be known about a route to take a shortcut or a detour? Cornell and Heth (Chapter 5) also suggest that some more cross-cultural research is needed, along with historical investigation into past navigation techniques.

Lloyd (see Chapter 6) suggests a number of issues in cartography that are in need of examination. Although there has been a number of studies which have sought to determine how spatial knowledge learnt from map presentations is stored in long-term memory, as yet, we have no clear models. Moreover, we do not know how map information displayed using animation techniques and alternative forms of map presentation such as cartograms are processed and remembered. Lloyd suggests that one strategy that might help guide future studies is to forge a link with visual information processing research. Whilst this will no doubt be an important link for comprehending sighted people's understanding of maps, it might have limited utility in comprehending how people with visual impairments learn, remember, and utilize spatial information derived from tactile maps.

Virtual environments are a relatively recent phenomenon, but are likely to become common place over the coming years. Because they are navigated through in ways visually similar to, but differing in other ways from, real world environments (e.g., lack kinaesthetic feedback), they offer an interesting medium in which to study spatial knowledge acquisition. As documented by Péruch *et al.* (Chapter 7), although studies of virtual environments have started in earnest, the number of studies remains relatively small. Key questions that need to be addressed relate to the extent to which spatial knowledge is the same or differs from that acquired in the real world, and how virtual environments can be designed to facilitate wayfinding. The latter point is particularly important given that, at present, navigation through these environments often causes disorientation (Ruddle *et al.*, 1997; Richardson *et al.*, in press).

Processes of integration

As most contributors to this volume have noted, in order to build a comprehensive theory of the process of cognitive mapping, it is necessary to examine processes of integration of spatial information. Studies are needed that explore how new spatial information is integrated with existing spatial knowledge, how spatial knowledge previously learned becomes distorted or forgotten, and how spatial information gained from travelling various routes through an environment is linked together. In particular, several authors noted the need to examine how spatial information from different media, such as direct experience, map representations, virtual reality, film/animation, text, and verbal dialogue is integrated both in storage and use (Montello and Freundschuh, 1995; Freundschuh and Taylor, 1999). At present, few studies have examined processes of integration, and our understanding, therefore, is mainly conceptual. It is clear, however, that these processes are central to the process of spatial learning and spatial thought. For a more complete understanding, detailed empirical evidence is needed to support what, at present, largely remain hypothetical explanations.

Spatio-temporal effects

SCALE

A number of studies provide empirical evidence that enables the distinction of 'manipulable space', 'environmental space' and 'map space' as 'cognitively different' (see Chapter 8). Further research is needed to explore these spaces, and to explore other kinds of spaces such as panoramic and geographic spaces, and other kinds of spaces that have yet to be identified (e.g., virtual reality). In addition to this work, systematic studies of the cognitive factors and abilities that are used to understand changes in scale and scale transformations is needed. We understand very little about how small-scale representations of a space are related to the larger-size space, or how and if multiple representations are used to cognitively link these spaces. In relation to understanding scale changes, longitudinal studies are needed that detail the development of scale understanding, from childhood to late life. These studies of 'life-long' learning will provide critical insights about benchmarks for learning, and possibily the un-learning of scale and spatial knowledge. An understanding of scale and size of space, we believe has important implications to the development of spatial tools, such as maps and GIS, and to the development of educational curriculums.

ENVIRONMENTAL DESIGN

Several studies have shown that environmental features can influence the rate, success and accuracy of spatial learning. For example, Canter (1977),

Cohen *et al.* (1978), and Herman *et al.* (1983) have all found barrier effects on distance judgements, where those locations separated by a barrier (e.g., a river) were judged to be further apart. Similarly, city layout, structure, and size have all been found to affect the cognitive knowledge of urban areas. For example, Antes *et al.* (1988) found that cities with a regular layout made the city more 'legible', and Sadalla and Staplin (1980) and Kahl *et al.* (1984) reported that the greater the number of turns or inter-sections along a route the greater the chance that cognitive distance estimates are overestimated. As Cornell and Heth (Chapter 5) detail, however, a more sustained analysis of role of intermediary design factors is needed, along with an analysis of the effects of specific interventions designed to make an environment easier to navigate through (e.g., colour coding, the role of signage). Determining these effects are important because, as discussed below, people interact with these environments, not with controlled labo-ratories where such factors are often missing.

TIME

One factor that has received little attention to date has been time. We live and move through a time–space continuum. Temporality, then, may have an important role to play in the spatial cognition on an environment in two respects (Kitchin and Blades, forthcoming). First, time relates to *when*; the specific time in which a place is experienced. Places experienced at different times might lead to different spatial understanding. For example, at night, when visual stimuli are darkened, the effectiveness of spatial cues may be lessened; during a rush hour cognitive attention may be stretched and engaged in other tasks such as obstacle avoidance rather than spatial layout. Second, the time of travel – the speed of *mobility* – may lead to differing spatial understandings. For example, driving quickly through an area may provide little time to remember spatial information; walking slowly may allow time to gaze and note the relative locations of objects in the environment. It is our contention that the interrelationship between space, time, and cognition needs to be fully examined. Related to time, of course, is the age at which spatial information is acquired and which spatial concepts are understood (see Chapters 8, 9 and 10). Acquisition of information at different ages will impact what spatial information is learned and what spatial concepts will be understood, as well as what spatial knowledge will be acquired and what knowledge will be forgotten. Also related to time is the understanding of change in spatial phenomena. Research in the use of animation to illustrate temporal changes in spatial phenomena are needed, as well as the development and integration of animation techno-logies in GIS.

Spatial language

We have only recently started to address questions concerning the role of language in spatial cognition. As a consequence, as shown in the chapters by Tversky (Chapter 3) and Taylor (Chapter 11), language-orientated research is much needed. Both suggest that research should centre on analysing spatial descriptions, such as route and survey descriptions, using these to investigate basic questions concerning knowledge form and structure. It seems to us that there are a number of key concerns that need to be addressed. First, attention needs to be directed at understanding how verbal and written spatial descriptions are learnt and processed, how such processing is similar to or differs from knowledge gained from different media (direct experience, maps), and how this spatial information is integrated into existing knowledge (Freundschuh and Mercer, 1995). Second, attention needs to focus on how spatial language is used as medium for communicating spatial relations, both verbally and written. Third, research needs to address how spatial language develops across the life-span, and to investigate whether there are cultural, gender-related or individual differences in language style and form, and the nature and bases of any differences.

Linking spatial thought and spatial behaviour

A number of contributors to this volume (e.g., Tversky, Gärling and Golledge, Cornell and Heth) argue that much more work is needed to try and understand the relationship between spatial knowledge and spatial behaviour (action); of the processes that are employed in guiding spatial choice and decision-making. Gärling and Golledge correctly assert that to date we have been lax in investigating this process, and yet it is a key reason for studying cognitive map knowledge. They argue that a sustained, experimental approach using small- and large-scale, real world environments is needed in order to ascertain the mechanisms by which spatial knowledge (however complete) is used in making spatial decisions. Central to this study is a focus on the *process* of decision-making and an appreciation of theoretical models of decision-making *per se*.

Development across the lifespan

As the chapters by Uttal and Tan (Chapter 9) and Kirasic (Chapter 10) illustrate, research concerning the development of spatial knowledge across the life-span has been unevenly distributed. At present, research has tended to focus on the development of spatial knowledge and abilities in childhood, and in particular infancy. Research relating to older adults is lacking despite the evidence to suggest that an understanding may have many potential applications given an ageing society with potentially deteriorating spatial skills. It is therefore no surprise that we contend that a significant area for future research must be an examination of spatial abilities in later life. In particular, research needs to establish the extent to which there is deterioration of spatial abilities in older age, identify the processes that underlie this deterioration, and determine ways to implement findings into environmental design.

The urgent need for basic research concerning spatial knowledge and abilities of older adults, does not however, negate the fact that fundamental questions concerning childhood development still need to be addressed further. As Uttal and Tan detail (see Chapter 9), future research in the development of cognitive mapping will include studies exploring scale effects (see also Chapter 8), new technologies such as virtual reality (see also Chapters 7 and 8) and tracking technologies (GPS), and the causes of development. GPS (global positioning systems) will enable the tracking of children's movements, making it possible to track changes in range of exploration that are related to age and experience. Research exploring the 'mechanisms of change' in the development in cognitive mapping will help researchers understand the specific factors and/or benchmark events that result in significant changes in cognitive mapping and spatial abilities.

Individual differences

In general, research to date has concentrated on identifying coherent trends in data. As such, analysis has generally taken place at the aggregate/ group level. As several contributors to this volume argue however, there is now a need to explore the differences between individual performances. These differences as Kirasic (Chapter 10) notes are predicated upon factors such as individual information processing abilities, personality variables, physical capabilities, and neurological states. The relative importance of each of these factors needs to be established, especially as within group differences are often larger than between group differences, and as noted by Kitchin (see Chapter 2), that group-based aggregations often lead to weak internal validity due to ecological fallacy (also see Kitchin and Fotheringham 1997). Until recently, calculating individual differences was relatively unviable due to the computation effort of data preparation and analysis. The development of powerful computers that perform complex spatial analysis enable the analysis of individual data in a matter of seconds rather than hours or days. This means it is now possible to perform disaggregate analysis, on individual data sets, establishing possible reasons for within-group differences.

Comparators

Although it is clear that substantially more research is needed in regard to examining and understanding individual differences it is also the case that further comparator work is also necessary. In particular, research is needed to compare performances across cultures, across species, between sexes, and between sighted and non-sighted populations.

As yet, there have been few attempts to conduct cross-cultural, cognitive mapping research. As such, the influence of cultural environments (e.g., urban form, street layout), linguistic styles, and other cultural factors upon spatial knowledge and tasks are unknown. As noted by Cornell and Heth (see Chapter 5), those studies which have been undertaken highlight differences in methods of navigation based upon surrounding environments.

Similarly, there has been little research that has compared spatial knowledge and spatial behaviour across species. It is clear that many animals possess complex spatial abilities that allow them to travel great distances accurately. The mechanics of these abilities and how they compare to human abilities is largely unknown. Comparison studies will not only throw light on animal behaviour, but may reveal clues as to how we as a species comprehend and use spatial knowledge. As noted below, the methodologies used to produce comparable data needs to be thoroughly investigated but it seems like that neuro-psychological work might be a fruitful venture.

As Chapter 12 highlighted there has been considerable research attention applied to the issue of gender differences in spatial abilities. These studies have produced findings that indicate conflicting conclusions. Some studies have found significant differences between males and females on a number of tests. Other studies have found no such differences. Hypotheses to explain the differences between the sexes centre on a number of themes such as physiology and hormone levels, socio-cultural factors (e.g., early childhood training and expectations, parental and institutional expectations, stereotyping and experience), and abilities to cope with the task presented rather than abilities *per se* (Kitchin, 1996a). The viability and integrity of these hypotheses still needs to be established. As Self and Golledge (1994) have acknowledged elsewhere, determining and understanding differences is important for activities such as teaching to ensure equivalence of knowledge.

The final set of comparator work that needs to be addressed is that of sighted versus non-/partially-sighted. To date research comparing sighted and blind populations has been limited in number and scope. As Ungar (see Chapter 13) notes, several contrary findings have been reported, each attributing different levels of spatial ability to non-sighted populations. In the main, research has largely been confined to laboratory testing, but in recent years some studies have investigated spatial knowledge and behaviour in large-scale, real environments. This research tends to suggest that the spatial abilities of visually impaired people have been underestimated. Ungar suggests that a concerted research programme is needed to address both abilities to learn and interact with large-scale environments, and abilities to learn and use knowledge derived from spatial communication devices such as maps and personal guidance systems. In addition, Ungar suggests that research should focus on a number of key themes, namely early experiences, particularly the role of care-givers and intervention strategies, an examination of varying degrees of visual impairment, the effect of new spatial technologies on spatial abilities, the influence of environmental design, and what insights can be gained from neuro-psychological work.

Knitting specific theories

> . . . we operate on two levels, both as model builders concerned with a particular aspect of our subject and as students of our entire subject. For some, there is but one level: their intellectual curiosity has shrunk to the size of a specialty.
>
> Papageorgiou (1982: 346)

So far in this section we have discussed agendas relating to specific aspects of spatial knowledge. Equally important, however, is the theoretical knitting of these ideas into a conceptual whole. To date there has been little attempt to knit together specific theoretical explanations beyond generalized conceptual models of spatial thought and spatial behaviour (see Kitchin 1996b, for review). These generalized models tend to be highly abstract and lacking in specific details, and are generally used to provide a conceptual frame in which to guide empirical research. Indeed, one of the main criticisms of cognitive mapping research to date has been that while there has been no shortage of empirical studies, these have been motivated by hypotheses that are too limited to be of general applicability, or too general to have been meaningful hypotheses in the first place (Allen, 1985). As such, present theories are often too specific to relate to cognitive mapping in general (e.g., structure, form, learning strategy), or too vague to give rise to testable hypotheses (e.g., environment-behaviour interaction schemata). Moreover, Golledge *et al.* (1985) contend that in many cases cognitive mapping theories represent general positions rather than formal models, and that empirical studies are often not explicitly tied to formal models. As a consequence, a major hole in our understanding of the process of cognitive mapping is a comprehension of this process in entirety. A major theoretical project for the future then is to fully explore both conceptually and empirically the process of cognitive mapping as a system, integrating specific theories into a conceptual whole.

Methodological

As detailed in Chapter 2, whilst methodology has become more sophisticated there is still a need to improve methodological validity and integrity. This can be achieved in four main ways. First, there needs to be more research conducted on the actual process of research itself. It is now quite clear that the process of data generation has significant implications to the findings of a study. For example, a technique can introduce methodological bias through spatial and location cueing (providing information to the respondent such as a spatial framework, e.g., part of a map, or a list of places to locate). Similarly, aggregation introduces effects through the removal of variance. The extent to which these biases influence the findings needs to be carefully established so that there effects can be compensated for. Until this is done, the integrity and validity of studies remains less robust.

Related to establishing the integrity of traditional research methodologies, is determining the applicability of using particular techniques to measure spatial knowledge acquired from different media. Whilst it seems intuitive that techniques which work well in a particular context such as the real world environment will work equally well in virtual environments, this needs to be established (see Chapter 8). Virtual environments, at present, do differ substantially in form from real world environments. Moreover, whilst there is clearly more need for more cross-cultural and cross-species research, the methodological integrity of such studies needs to be established. The danger, in relation to research on cross-cultural comparisons, is that results may differ not because of differences in knowledge or cognitive processes but because of cultural familiarity with the media of data collection. In relation to cross-species work there is always a danger of placing an anthropomorphic interpretation of findings.

Second, empirical studies need to adopt research strategies that allow some level of construct and convergent validity to be established. Construct validity refers to the extent to which a methodology is measuring what it is supposed to, and convergent validity the extent to which two methodologies designed to measure the same phenomenon produce similar findings. Establishing construct and convergent validity is important because it provides a 'natural' way to determine the integrity of the findings through a process of cross-checking. One method to achieve this is to use multiple strategies of data collection and analysis, comparing the results from different strategies to determine equivalence of findings. In this manner the internal and external validity of a study can be verified.

Third, new methods of enquiry and specific techniques of data generation and analysis need to be utilized and evaluated. Several of this volume's contributors have argued that methodological developments in neuro-psychology, such as PET, could add significantly to our understanding of spatial knowledge. While there is a vast neuropsychological literature concerning the brain bases used during cognition, at present, theoretical and methodological developments within neuro-psychology have been slow to work their way into conventional cognitive mapping research. This in part is because neuro-psychologists have mainly focused their attention on the neurological bases of animal wayfinding, and in particular rats, but also because most psychologists and geographers are interested in what the neurons do rather than how they physically do it. Clearly the two, however, are related and there may be important insights to be learnt from marrying these two separate areas of research. This will be a difficult and lengthy task, but the development of connectionist models of neural networks may offer one path forward (see Chapter 6).

In addition, it has been speculated that qualitative methodologies, both those used within a scientific frame and those that are more interpretative in nature, may have significant utility as techniques for understanding spatial thought and spatial behaviour. Within a scientific frame, qualitative methodologies, such as talk-aloud protocols, will allow insights into the role of spatial language and, as a media of producing spatial products, other facets of spatial knowledge. Interpretative techniques, on the other hand, will allow an investigation of the role of value systems, situational context and socio-cultural factors relating to interrelationship between individual and society which is not easily captured using closed quantitative techniques such as questionnaires.

Fourth, we would suggest that more research needs to be conducted within natural settings rather than artificial laboratory space (see Chapter 8). Spatial behaviour takes place in complex, natural settings, often full of other people, not in highly controlled spaces that are devoid of life. Studies which take place in the orderedness of a laboratory may well suffer from problems of ecological validity. In the process of trying to create a controlled space to measure specific processes of cognitive mapping, the environment becomes something different to that normally experienced. What we are therefore measuring may differ substantially from what actually occurs when learning or interacting within a real world environment. Furthermore, there is a need to recognize and examine the situational context of spatial learning and spatial behaviour. These processes do not take place in a vacuum, but within a context that impinges upon their functioning. For example, there is evidence to show that rates of spatial learning are affected by the condition under which a location is experienced: passive explorers, for instance, show lower levels of spatial learning than active explorers (Feldman and Acredolo, 1979; Herman, 1980). A focus for study then is a detailed examination of the spatial (laboratory or natural setting) and situational context (condition) of learning and interacting in different environments.

Application

As noted in Chapter 1 particularly, as well as in subsequent chapters, cognitive mapping research tends to be conceptual in nature, despite repeated rhetoric concerning its potential applied worth. So far little attempt has been made to convert research findings into formal guidelines for application to specific issues. This to our reckoning is one of the biggest failings of cognitive mapping research to date, and one that needs significant attention.

Planning

One of the most consistent arguments for cognitive mapping research by geographers and planners is the potential for findings to be used to create environments that are easier to wayfind in and more pleasant to interact with. As Gärling and Golledge (1989: 203) and Lynch (1976: xi) stated:

> knowledge gained about perceptual-cognitive processes may improve the quality of human environments through policy, planning, and design, to the extent that it tells us how to plan and design environments that do not interfere with the proper functioning of these processes.

> . . . [we] can better plan, design and manage the environment for and with people if we know how they image the world.

There is little evidence to suggest that the findings of cognitive mapping researchers are being used to improve urban planning. Part of the reason for this is that researchers have been poor at communicating their findings and their potential implications beyond academic journals. These journals are not widely read by city planners or those in a position to make concrete changes on the ground. As a consequence, there is a need to produce a set of formal guidelines that explicitly detail how findings to date can translate into planning good practice. These guidelines need to be inclusive in nature so that they take account of different groups within society such as children, older people, and people with disabilities.

Wayfinding, spatial searches and teaching

A large proportion of cognitive mapping research has focused on examining how people learn routes through a city. To date, however, little research has considered how best to teach people more effective and efficient strategies of route finding; how best to cope with feelings of disorientation or of being lost; or the most effective and efficient ways of guiding people through an environment (e.g. sign type/location). One exception to this claim is the work by Streeter *et al.* (1985). In their study, Streeter *et al.* demonstrated that subjects using tape recorded instructions for navigating between two unknown locations had shorter travel times and made fewer navigation errors than subjects who relied on unmodified road maps, modified road maps, and modified road maps and recorded instructions. All in all, we know of no formal guidelines that advise on these issues beyond highly specialized courses given to individuals taking part in particular sports such as orienteering or mountain walking, or orientation and mobility training for people with visual impairments. In these cases, the methods taught have been long established and fail to draw from the insights of cognitive mapping research. Traditional means of aiding wayfinding such wayfinding such as signage are poorly understood and work more from commonsense practice than proven efficient utility.

In addition, it has been hypothesized that cognitive mapping research can be of potential benefit to those involved in professional spatial searches (e.g., police, mountain rescue), by providing an indication of likely patterns of spatial behaviour of those being searched for. In relation to the police it is believed that cognitive mapping research can highlight the likely spatial behaviours of criminals, allowing police to predict approximate locations of the offender's residence and future targets. In relation to search-and-rescue it is hypothesized that research will identify likely movements upon becoming lost or going missing, allowing searchers to narrow the field and time of their search. In some cases, cognitive mapping researchers are now working with professional searchers and are increasing the effectiveness of searchers (e.g., Canter and Larkin, 1993, Cornell *et al.*, 1996), but these links needs to be further established.

Despite rhetoric of how cognitive mapping research could improve the teaching of geographic concepts, there have been no specific guidelines of how to instigate such improvements, instil more effective strategies of spatial learning amongst pupils, or how to design geographic media so that they are more easily understood. As such, such guidelines need to be formulated and their worth effectively marketed to the teaching profession.

Geographic media

Many recent projects have investigated how people learn and use geographic information derived through secondary sources. In part, the research is motivated by a desire to improve the efficiency and effectiveness of communication through geographic media. As yet, however, there have been few guidelines as to how to translate theory into practice. As a consequence, most geographic media are designed and implemented with little reference to cognitive theory. This is now slowly beginning to change, particularly in reference to cartography and GIS, where a number of researchers based mainly in North America have published widely on cognitive cartography. Consequently, cognitive theory is now starting to form the underlying framework for standard texts on map design (e.g., MacEachren, 1995). In relation to GIS, substantial work has investigated the notion of naive GISs for use by non-experts (Mark *et al.*, 1997). However, this work has yet to have a significant impact on the systems produced by major GIS software companies such as Environmental Systems Research Institute (ESRI). It is a similar story for other geographic media such as orientation and navigation aids, in-car navigation systems and virtual reality systems where developments are, at present, technology and ideas driven with little consideration of cognitive theory. Again, there is a need for those engaged in empirical studies of cognitive mapping to try and translate their findings into formal guidelines for use by practioners.

Conclusion

Cognitive mapping research seems set to flourish as a multi- and hopefully interdisciplinary endeavour. The agenda outlined in this final chapter, if followed, will provide substantial additional insights into how people learn, store, process, and apply spatial knowledge relating to the environment that surrounds them. Whilst some of these insights will be achieved if the empirical studies are undertaken and theoretically contextualized within a disciplinary isolation, it is our contention that much more progress will be gained from interdisciplinary collaboration. Such collaboration will force many of us to re-evaluate our ideas, to explore alternative propositions, and push back the boundaries of study by challenging us to move beyond disciplinary-rooted, preconceived notions of theory and practice. This means that central to any future agenda of cognitive mapping research must be the development of appropriate integrative frameworks for study (Gärling *et al.* 1991). Indeed, Evans and Gärling (1991) contend that the integration of disciplinary paradigms (e.g., environmental psychology and behavioural geography) may be a fruitful venture because it will force an analysis that reveals points of convergence and divergence among topics of scholarly inquiry. They argue that integration might help to illuminate correct and incorrect models and hypotheses, and to shed constraining or incorrect paradigmatic restrictions. As Hanson (1983: 35) argues:

> only through the process of communication among divergent points of view, will any semblance of convergence ever be achieved or maintained; through discourse the bits and pieces can be fitted into larger structures, and some degree of order emerges from the mess. . . . At the heart of this process of change is communication.

As the multidisciplinary make-up of the contributing authors, and the bibliography of each chapter illustrates, researchers within specific disciplines have much to learn from colleagues in other disciplines. This need for interdisciplinary research has become readily apparent to those researchers fortunate enough to attend the Cognitive Mapping Symposium in Fort Worth in 1997 and the symposiums sponsored by the National Center for Geographic Information and Analysis. Indeed many of these attendees are now engaged in cross-disciplinary projects (including ourselves). We would urge all those engaged in cognitive mapping research to similarly explore research beyond their disciplinary boundaries. Such exploration will maintain and expand the vibrancy of the field, and advance methodological validity and theoretical integrity.

References

Allen, G. (1985) Strengthening weak links in the study of the development of macrospatial cognition. In Cohen, R. (ed.), *The Development of Spatial Cognition*. Hillsdale, NJ: Erlbaum, pp. 301–21.

Antes, J.R., McBridge, R.B. and Collins, J.D. (1988) The effect of a new city traffic route on the cognitive maps of its residents. *Environment and Behavior*, 20, 75–91.

Canter, D. (1977) *The Psychology of Place*. London: Architectural.

—— and Larkin, P. (1993) The environmental range of serial rapists. *Journal of Environmental Psychology*, 13, 63–9.

Cohen, R., Baldwin, L.M. and Sherman, R.C. (1978) Cognitive maps of a naturalistic setting. *Child Development*, 49, 1216–18.

Cornell, E.H., Heth, C.D., Kneubuhler, Y. and Sehgal, S. (1996) Serial position effects in children's route reversal errors. *Applied Cognitive Psychology*, 10, 301–326.

Evans, G.W. and Gärling, T. (1991) Environment, cognition and action: the need for integration, in Gärling, T. and Evans, G.W. (eds), *Environment, Cognition and Action: An Integrated Approach*. New York: Oxford University Press, pp. 3–13.

Feldman, A. and Acredolo, L. (1979) The effect of active vs passive exploration on memory for spatial location in children. *Child Development*, 50, 698–704.

Freundschuh, S.M. and Mercer, D. (1995) Spatial cognitive representations of story worlds acquired from maps and narrative. *Geographical Systems*, 2, 217–33.

—— and Taylor, H. (1999) Multiple modalities and multiple frames of reference for spatial knowledge, Varenius Research Initiative, Panel on Cognitive Models of Geographic Space, National Centre for Geographic Information Systems, URL: http://www.ncgia.ucsb.edu/varenius/initiatives/ncgia.html.

Gärling, T. and Evans, G.W. (eds), (1991) *Environment, Cognition and Action – An Integrated Approach*. New York: Oxford University Press.

—— and Golledge, R.G. (1989) Environmental perception and cognition. In Zube, E. and Moore, G. (eds), *Advances in Environmental Behaviour and Design 2*. New York: Plenum Press, pp. 203–36.

——, Lindberg, E. Torell, G. and Evans, G.W. (1991) From environmental to ecological cognition. In Gärling, T. and Evans, G.W. (eds), *Environment, Cognition and Action – An Integrated Approach*, New York: Oxford University Press, pp. 335–44.

Golledge, R.G., Smith, T.R., Pellegrino, J.W. Doherty, S. and Marshall, S.P. (1985) A conceptual model and empirical analysis of children's acquisition of spatial knowledge. *Journal of Environmental Psychology*, 5, 125–52.

Hanson, S. (1983) The world is not a stone garden. *Geographical Analysis*, 15, 33–5.

Herman, J.F. (1980) Children's cognitive maps of large-scale spaces: effects of exploration, direction and repeated experience. *Journal of Experimental Child Psychology* 29, 126–43.

——, Norton, L.M. and Roth, S.F. (1983) Children and adults distance estimations in a large-scale environment: effects of time and clutter. *Journal of Experimental Child Psychology*, 36, 453–70.

Kahl, H.B., Herman, J.F. and Klein, C.A (1984) Distance distortions in cognitive maps: an examination of the information storage model. *Journal of Experimental Child Psychology*, 38, 134–46.

Kitchin, R.M. (1996a) Are there sex differences in geographic knowledge and understanding? *The Geographical Journal*, 162: 273–86.

—— (1996b) Increasing the integrity of cognitive mapping research: appraising conceptual schemata of environment-behaviour interaction. *Progress in Human Geography*, 20, 56–84.

—— and Blades, M. (forthcoming) *The Cognition of Geographic Space*. Baltimore, MD: Johns Hopkins.

—— and Fotheringham, A.S. (1997) Aggregation issues in cognitive mapping research. *Professional Geographer*, 49, 269–80.

Lynch, K. (1976) Preface. In: Moore, G. and Golledge, R.G. (eds), *Environmental Knowing*, Stroudsberg, PA: Dowden, Hutchinson and Ross.

MacEachren, A. (1995) *How Maps Work: Representation, Visualization and Design*, New York: Guildford.

Mark, D.M., Egenhofer, M.J. and Hornsby, K. (1997) *Formal Models of Commonsense Geographic Worlds: Report on the Specialist Meeting of Research Initiative 21*, National Centre for Geographic Information and Analysis, University of Maine, Orono.

Montello, D.R. and Freundschuh, S.M. (1995) Sources of spatial knowledge and their implications for GIS: An introduction, *Geographical Systems, Issue on Spatial Cognitive Models*, 2, 169–76.

Papageorgio, Y.Y. (1982) Some thoughts about theory in the social sciences. *Geographical Analysis*, 14, 340–6.

Richardson, A.E., Montello, D.R. and Hegarty, M. (in press) Spatial knowledge acquisition from maps, and from navigation in real and virtual environments. *Memory and Cognition*.

Ruddle, R.A., Payne, S.J. and Jones, D.M. (1997) Navigating buildings in 'desktop' virtual environments: Experimental investigations using extended navigational experience. *Journal of Experimental Psychology: Applied*, 3, 143–59.

Sadalla, E.K. and Staplin, L.J. (1980) The perception of traversed distance – Intersections. *Environment and Behavior*, 12, 167–82.

Self, C.M. and Golledge, R.G. (1994) Sex-related differences in spatial ability: what every geography educator should know. *Journal of Geography*, 93, 234–43.

Streeter, L.A., Vitello, D. and Wonsiewicz, S.A. (1985) How to tell people where to go. *International Journal of Man-Machine Studies*, 22, 549–62.

Index